Carbon Sequestration

Edited by Suriyanarayanan Sarvajayakesavalu and Kannan Karthikeyan

Published in London, United Kingdom

Carbon Sequestration
http://dx.doi.org/10.5772/intechopen.97965
Edited by Suriyanarayanan Sarvajayakesavalu and Kannan Karthikeyan

Contributors
Juliana Feitosa Felizzola, Ricardo de Oliveira Figueiredo, Wenceslau Geraldes Teixeira, Bruno Carneiro, Umi Asli, Mohd N.H. Sarjuni, Siti A.M. Dolit, Aidee K. Khamis, Nazrin Abd-Aziz, Nur R. Azman, Donald L. Rockwood, Martin F. Ellis, Kyle W. Fabbro, Monica Ozores-Hampton, Amir Varshovi, Yen A. Sokama-Neuyam, Muhammad A. M. Yusof, Shadrack K. Owusu, Yongcun Feng, Shuai Zhang, Luisa Epimenia Moreno Reyes, Jorge Eliecer Mariño Martínez

First published in London, United Kingdom, 2022 by IntechOpen
IntechOpen is the global imprint of INTECHOPEN LIMITED, registered in England and Wales, registration number: 11086078, 5 Princes Gate Court, London, SW7 2QJ, United Kingdom

British Library Cataloguing-in-Publication Data
A catalogue record for this book is available from the British Library

Additional hard and PDF copies can be obtained from orders@intechopen.com

Carbon Sequestration
Edited by Suriyanarayanan Sarvajayakesavalu and Kannan Karthikeyan
p. cm.
Print ISBN 978-1-80355-687-1
Online ISBN 978-1-80355-688-8
eBook (PDF) ISBN 978-1-80355-689-5

Meet the editors

Prof. Dr. S. Suriyanarayanan, MSc, MPhil, Ph.D., is the deputy director of research, at Vinayaka Mission's Research Foundation Deemed University, India. Prior to this position, he was a post-doctoral fellow at the University of Turin, Italy, and project coordinator, faculty, and head (I/c) at the Department of Water and Health, JSS Academy of Higher Education, and Research (JSS AHER), Mysore, India. He also served as a science officer for the Scientific Committee on Problems of the Environment (SCOPE) and the Research Center for Eco-Environmental Sciences (RCEES), Beijing from April 2015 to March 2017. He is also a nodal officer for SCOPE India activities under JSS AHER. Dr. Suriyanarayanan has research experience in environmental monitoring, radiation ecology, and environmental microbiology. He is a recipient of the Young Scientist research grant award from the Science and Engineering Research Board (SERB), Department of Science and Technology (DST), Government of India. He also received a visiting scientist fellowship from the Chinese Academy of Sciences in 2016–2017. He has been a chairperson, participant, and presenter at many national and international conferences as well as an editorial board member for peer-reviewed journals. Dr. Suriyanarayanan has more than forty published research articles to his credit and is an associate editor for Environmental Development.

Dr. K. Karthikeyan is a senior scientist at Gujarat Institute of Desert Ecology, Bhuj, India. He obtained a Ph.D. in Environmental Sciences from Bharathiar University, Coimbatore, India. He has more than sixteen years of work experience in environmental monitoring, bioprospecting, bioremediation, and industrial auditing. He has published thirty-five research papers and two book chapters. He was accredited by the National Accreditation Board for Education and Training, Quality Council of India (NABET-QCI), New Delhi, as a Functional Area Expert in Water Pollution Monitoring, Prevention and Control. He is ISO-certified in laboratory management training and internal auditing. Dr. Karthikeyan is a recipient of the 2009 Best Environmentalist Award from Bharathiar University.

Contents

Preface

Increased quantities of carbon dioxide (CO_2) and other greenhouse gases in the atmosphere are causing global warming. Capturing and storing CO_2 is critical for meeting global climate change goals. The process of capturing and storing atmospheric CO_2 is called carbon sequestration. Through this process, atmospheric CO_2 is captured by trees, plants, and other vegetation via photosynthesis and stored in biomass and soils. Enhanced agricultural practices, reforestation, and urban management could increase carbon sequestration capacity.

This book provides an overview of the recent developments in carbon sequestration, which plays an important role in climate change mitigation. It consists of six chapters that discuss carbon capture, storage, utilization, chemistry, and geomechanical aspects. This book is a useful resource for students, researchers, decision-makers, chemical and powerplant engineers, geological and environmental engineers, and those who work in the industry.

Suriyanarayanan Sarvajayakesavalu
Professor and Deputy Director Research Vinayaka Missions
Kirupanandavariyar Arts and Science College,
Vinayaka Missions Research Foundation Deemed to be University,
Salem, India

SCOPE – India (*Scientific Committee* of *Problems of the Environment*),
JSS Academy of Higher Education and Research (JSSAHER),
Mysuru, Karnataka, India

Dr. Kannan Karthikeyan
Senior Scientist and Head,
Environmental Laboratory Division,
Gujarat Institute of Desert Ecology,
Bhuj, Dist: Kachchh, Gujarat, India

Section 1

Carbon Capture, Storage and Utlization

Chapter 1

Carbon Capture, Use and Storage (CCUS) as Enhanced Oil Recovery (EOR): Llanos Orientales Basin (Colombia)

Jorge Eliecer Mariño Martínez and Luisa Epimenia Moreno Reyes

Abstract

At present, it is essential to take actions to minimize CO_2 emissions into the atmosphere, and one way to use and exploit it is through the use of carbon dioxide for industry processes, such as enhanced oil recovery. By carrying out carbon capture, geological storage and improving oil recovery by applying the selection criteria to determine suitability for CO_2 sequestration in the Llanos Orientales Basin as a tool to improve hydrocarbon production. It would be a first step in the sustainable development of the extractive oil industry in Colombia, which is one of the participants in greenhouse gas releases. The above can be achieved through the application of CO_2–EOR processes, which are mainly miscible or immiscible methods. Finally, the importance of identifying existing CO_2 sources is highlighted, in order to carry out the application of a CO_2–EOR project.

Keywords: CO_2 emissions, carbon, enhanced oil recovery, capture, geological storage, Llanos basin, CCUS, EOR

1. Introduction

Global warming is an imminent threat to humanity, and it is related to CO_2 emissions in the atmosphere, which is mainly the product of the combustion of hydrocarbons (oil and gas), coal in power plants, and other industrial plants.

Considering the above, carbon capture use and geological storage (CCUS) is being considered as one of the methods to reduce greenhouse gas emissions, thus applying the methods of optimizing the operation of the different projects of enhanced oil recovery (EOR) by using carbon dioxide (CO_2), becoming the tool to improve efficiency and profitability in the production of the hydrocarbons sector. In this way, it contributes responsibly to sustainable development in energy projects [1].

The geological, geothermal, and hydraulics properties of the Eastern Llanos basin are favorable for the storage of CO_2. Considering the main criteria for the application of these projects, especially in the cases of mature deposits, it would become an applicable method to increase the production of hydrocarbons through improved recovery and thus increase their useful life.

2. Current situation of CO_2: EOR

Enhanced oil recovery is done by the use of CO_2 (CO_2–EOR). It is a kind of CCUS technology. This practical application was started and improved as early as 1972, directing to improve oil recovery through the introduction of CO_2 into oil reservoirs. (**Figure 1**) [2].

The current and active projects, where CO_2–EOR is applied, have been developed mostly in the United States (Permian Basin, Gulf Coast, and the Rockies (see **Figure 1.**)), and in Canada in better proportion with seven projects of injection of CO_2, plus another additional injection of acid gases; however, it is relevant to indicate that this method has also been practiced for several decades in Turkey and Hungary, and there are new projects in various stages of development in Asia, the Middle East, and the North Sea. On a pilot scale, there are also developments in China, Brazil (onshore – Miranga field), and Abu Dhabi.

As a consequence of the advantages of this method, as part of the KAPSARC data source, in 2020, there are 38 CO_2–EOR large-scale projects in different project life cycle stages [3], and according to International Energy Agency (IEA), the entire amount of CCUS programs in industry and fuel transformation increase to 19 in 2020 when the two Alberta Carbon Trunk Line programs in Canada started activity (**Figure 2**) [2].

According to the information updated in 2020, it was identified that there was a decrease in the U.S. of approximately 47% of CO_2 supplied for EOR projects compared

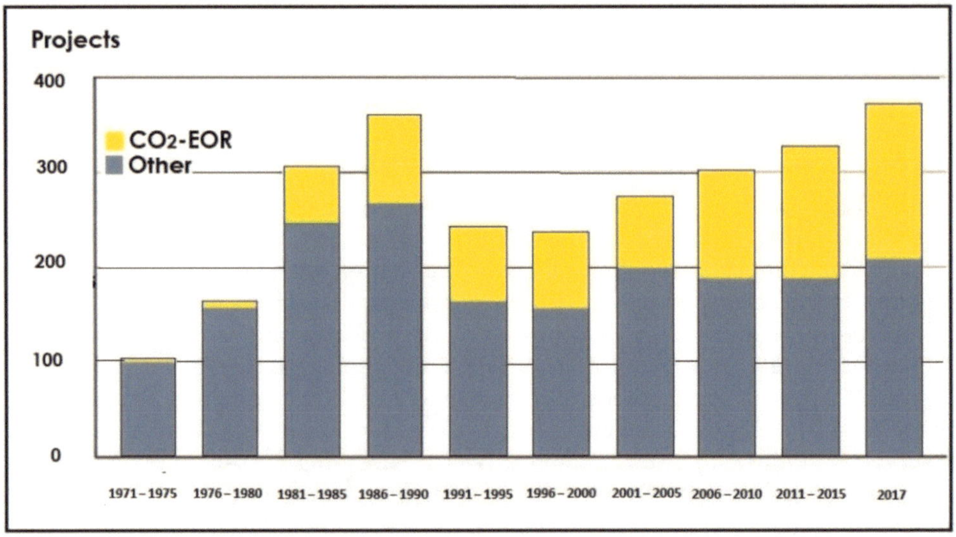

Figure 1.
Number of EOR projects in operation globally [2].

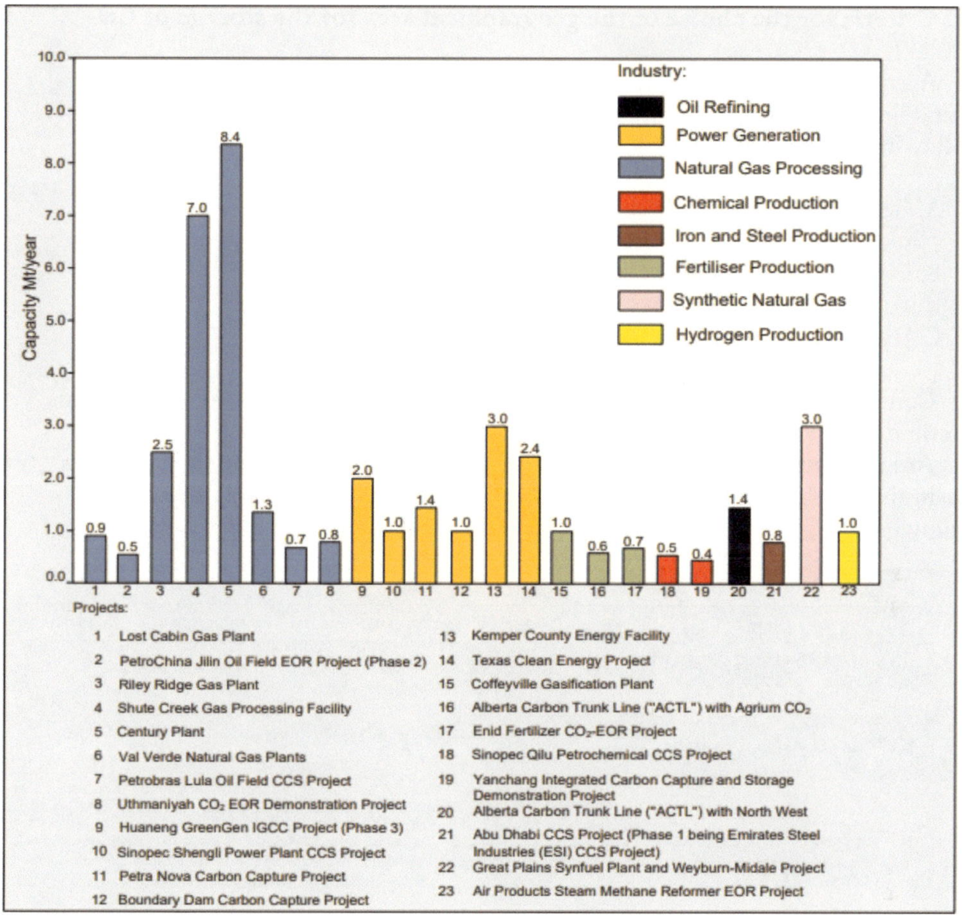

Figure 2.
Sizeable CO_2–EOR projects in different stages by applied industries [5].

to that supplied in 2019, with one of the possible causes being the decline in oil and gas prices. However, the increment in prices in 2021 may lead to an increase in oil production through EOR–CO_2 projects [4].

Figure 2 indicates that the Century CO_2–EOR plant in Texas is one of the projects with the most capacity per year. It was started in 2010 with a reduced capture capacity, but the volume was increased to its full capacity of 8.4 Mt./year in 2012 [5].

Also, the Shute Creek Gas processing installation in Wyoming, USA, has been handling natural gas from the LaBarge field beginning in 1986. Before improving it, H_2S was separated along with approximately 0.4 Mt./a of CO_2. In 2010, an expansion of the plant's capacity was finished, getting a capture capacity of 7 Mt./a of CO_2 [5].

Another project is the Val Verde Natural Gas Facility in Texas, USA. Currently, five different gas processing installations in the Val Verde area get about 1.3 Mt./yr. of CO_2 for use in EOR facilities at the Sharon Ridge oil field. The CO_2 concentration of the incoming gas stream at the Val Verde plant ranges from 25 to 50% [5].

Under evaluation phases, there are eight big CO_2–EOR projects. Nearly 63% of them are planned in China irrespective of the reality that tight continental geology and heavier oil are important factors [5].

2.1 Criteria for the choice of the geographical area for the storage of CO_2

According to Bachu [6], it is necessary to evaluate, at the basin scale, the suitability for CO_2 sequestration in sedimentary rocks, among which are the following aspects:

A. Geological criteria

B. Geothermal criteria

C. Hydrodynamic criteria

Considering the above, a roadmap is proposed for the implementation of this method in geological media [7]:

Assessment of suitability at the regional level, to determine the areas of a basin that could become suitable and the means of storage, which include the following

Basin geology and hydrostratigraphy	To identify specific areas and beds suitable for CO2 storage as a result of extensive confinement, and eliminate areas and strata where CO2 migration is possible at the local scale.
Basin hydrodynamics	To identify regions and flow systems proper for hydrodynamic capture.
Temperature and pressure distributions	To establish the physical state and density of the CO2 expelled under in situ conditions. This is particularly essential to avoid areas and strata where CO2 phase change and instability can lead to rapid migration and escape of CO2, and to establish storage capacity.
Oil and gas fields and their location	In the context of regional suitability, phase and density of CO2, for the use of CO2 in enhanced oil recovery (EOR) and storage in depleted fields.
Coal beds and their location	Depth, maturity, methane content, water quality, and pressure regimes, for the storage of CO2 in uneconomic coal layers and increased methane production.
Salt beds	Their location and pressure regime, for the storage of CO2 in caves.

Inventory of feasible sites for CO_2 storage, which initially requires the identification of the main sources of CO_2, followed by a local-level characterization considering the pressure and temperature *in situ*, as well as properties of hydrocarbons.

Safety of CO_2 storage operations, where it is secure that there will be no upward migration of CO_2 and leakage to other beds during or after introduction, which could occur through open geological faults and natural or man-induced fractures, irrespective of storage media.

Storage capacity, for EOR operations in mature stage oil and gas deposits, storage pore volume, grade of water penetration as a result of oil production, and CO_2 solubility are crucial components in judging storage capacity.

3. Benefits of lifecycle emissions on CO_2 reduction of CO_2: EOR projects

As reported by the International Energy Agency (IEA), the hydrocarbons industry is one of the world's pioneers in the development and deployment of CO_2 capture. Currently, most of the CO_2 injected in CO_2–EOR projects is produced from naturally occurring subsurface CO_2 deposits. It is clear that the use of natural sources does not provide any advantage in the emissions degree of the oil produced. In the United States, more than 70% of the CO_2 dispensed today for CO_2–EOR is from natural sources [8].

Nevertheless, there are some programs that use CO_2 captured from man-made sources for EOR—the Century and Petra Nova plants in Texas. These are two bigger installations, which are making the development of these projects more efficient and, thus, more profitable (**Table 1**).

Guaranteeing the integrity of CO_2 storage is also essential for confirming the emissions decreases. There are some path operators, which demonstrate the permanency of CO_2 storage, admitting sites with appropriate geology that gets CO_2, avoiding abandoned oil wells that could create a passage for CO_2 to get the surface (or checking that these are obstructed), and inserting observance and field surveillance to notice possible escape [8].

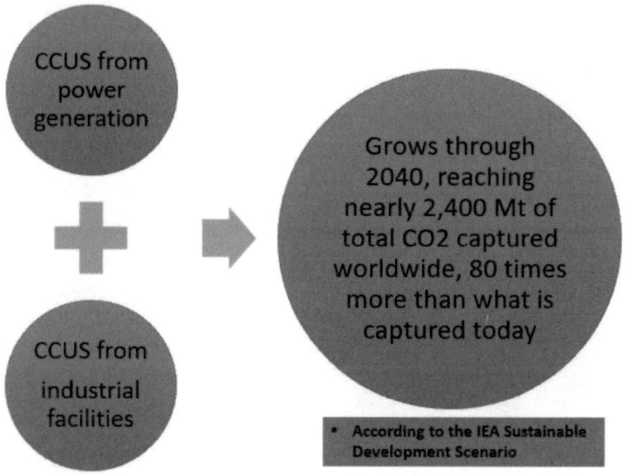

An additional potential advantage of CO_2–EOR is that it offers a lower-cost opportunity to position CCUS projects.

4. Enhance the oil recovery process using ccs

From a geological perspective, the traps can be structural and stratigraphic, and the EOR could be from abandoned or deleted reservoirs:

Projects of CO2–EOR	
The oil revenues produced or reduced total project costs and expand the amount of CO_2 stored per unit of investment.	Developing a series of such projects would help reduce CCUS costs more generally and could provide the catalyst for commercial-scale CCUS to eventually take off.

Table 1.
Benefits of lifecycle emissions on CO_2 reduction of CO_2–EOR projects [8].

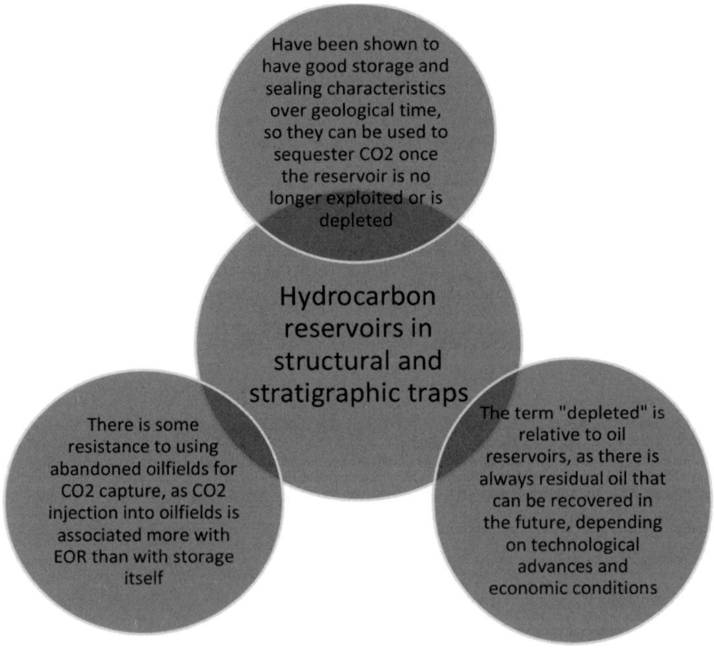

In addition to the information shown in **Table 2**, there are two types of CO_2–EOR based on the miscibility between CO_2 and the reservoir oil. The pressure at which miscibility occurs is defined as the minimum miscibility pressure (MMP) [9].

As part of the CO_2 – EOR operation, CO_2 is inserted into a stratum containing oil at high pressure. The displacement of oil by CO_2 injection is founded on the phase behavior of gas–oil mixtures, which are highly interdependent on the temperature, pressure, and constitution of the reservoir oil. Two main types of CO_2–EOR procedures are recognized [1].

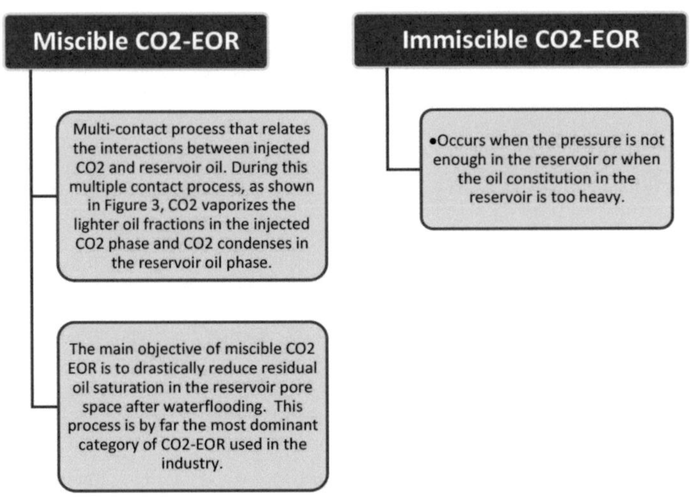

The main processes participating in the immiscible CO_2 floods are:

- Increased oil stage, as the oil gets soaked with CO_2.
- Viscosity decreases in the mixture of oil and CO_2.

Groups of EOR technologies:	Recovery of miscible gases
	Chemical flooding
	Thermal recovery
	Microbial flooding

Table 2.
EOR Technologies [9].

- Pumping of lighter hydrocarbons in the CO_2 phase. and
- Displacement of liquid by the increase of pressure.

This assemblage of mechanisms allows a part of the left-over oil in the reservoir to be still, moved and extracted (**Figure 3**).

CO_2 capture has been shown as one of the most successful EOR methods, especially in the United States, because it uses the available CO_2 deposits that are present in nature.

As can be identified in **Figure 4**, there are two sources of CO_2:

- The gas CO_2 is associated with the gas and oil pumping from which the oil company purifies it and transports it to the CO_2 liquefaction installation using pipelines, which is a free gas source.
- Liquid CO_2 is obtained from a chemical plant not so far from the project that is trucked to the CO_2 storage installation.

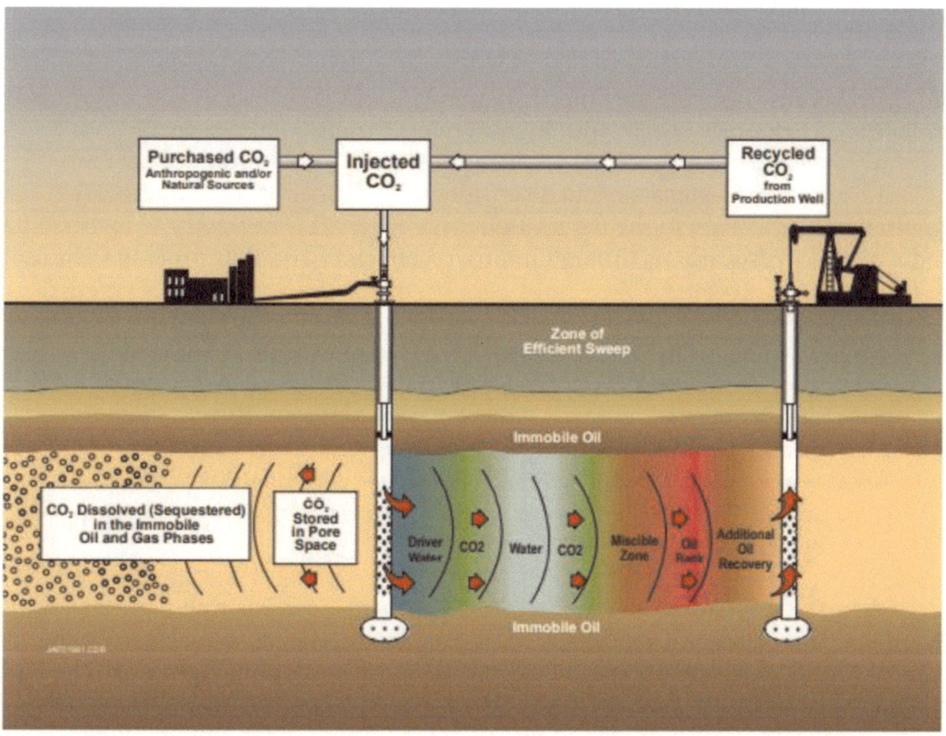

Figure 3.
Diagram presenting the miscible CO_2–EOR procedure [10].

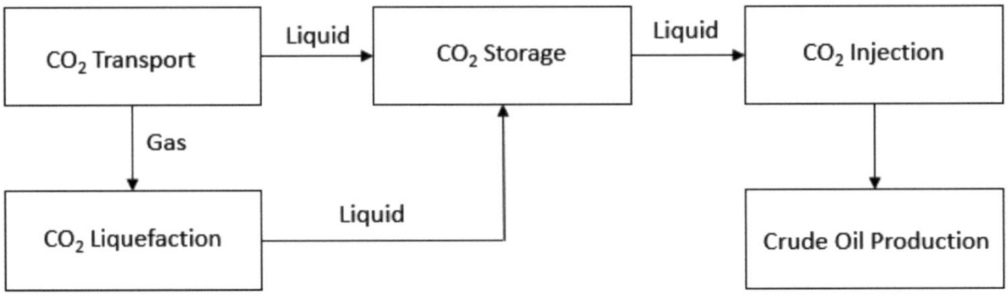

Figure 4.
Production flow diagram of the CO_2–EOR project [9].

As for the CO_2 extraction process, the first measure in CCS is to obtain it from other gaseous substances. CO_2 can be seized from natural gas by absorption, adsorption, chemical looping, or membrane gas detachment, or it can be captured from flue gas at big CO_2 location origin (coal power station and industrial processes units) by using one of the following three methods—pre-combustion capture, post-combustion capture, and oxy-combustion. Once obtained, the CO_2 is dehydrated and equipped for transport. It is transported by pipeline, truck, or ship in a supercritical state to the storehouse site, see **Figure 4** [5].

5. Possible applications of the method CCUS–EOR IN the EASTERN LLANOS basin (Colombia)

In general, the actual Eastern Llanos basin is a suitable basin for the storage of CO_2–EOR because it meets the criteria mentioned above, such as 1) natural gas and oil production, 2) extensive basin with hydrodynamic traps, and 3) several reservoir formations with good regional confinement. To this is added that oil fields are varied and extensive, and a considerable number of these hydrocarbon exploitation projects are in the maturation stage where it is necessary to increase the production of hydrocarbons through improved recovery and thus prolong their useful life (**Figure** 5).

> *"The Eastern Llanos Basin is the most prolific hydrocarbon basin in continental Colombia. The northern boundary of this basin is the Colombian–Venezuelan delimitation; to the south, the basin goes as far as the Macarena high, the Vaupés Arch and the Precambrian metamorphic rocks that outcrops to the south of the Guaviare river; the east limit is marked by the outcrops of Precambrian plutonic rocks of the Guyana Shield, and to the west, the basin is restricted by the frontal thrust system of the Eastern Cordillera"* (**Figure** 5) [11].

A schematic drawing cross section of the Eastern Llanos basin shows that the oil-related formations are of Cretaceous, Paleogene, and Neogene ages, between the Eastern Cordillera westward and the Brazilian craton eastward (**Figure 6**). The Eastern Llanos is a complete foreland basin that changed when the Andes were pushed to the east against the South American plate. The structural surroundings allow the geological formations in the basin that stayed mostly planar and undisturbed, fashioning them favorable for CO2 sequestration.

DOI: http://dx.doi.org/10.5772/intechopen.105949

Figure 5.
The Eastern Llanos basin located in the bigger Orinoco River basin in northern South America.

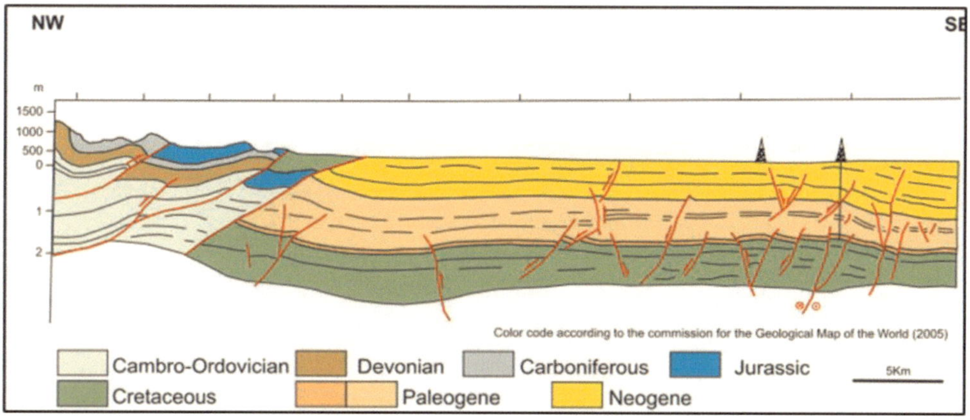

Figure 6.
Cross section of the foreland Eastern Llanos basin [11].

Basin development began in the Paleozoic with rifting. Clastic materials were sedimented over the Precambrian basement from the Triassic time to the Late Cretaceous. The basin was the oriental block of a major rift system. From the Maastrichtian to the Paleocene time, the Llanos basin developed into a foreland. During the Neogene, the basin has been a deposit of thickened molasse sediments. The source rocks are Cretaceous, and span from immature to marginally mature eastward of the frontal thrust. The main deposits are clastic units of the Late Cretaceous and Paleogene ages. Investigation of the individual elements of the migration arrangement inside the basin is complex because of the thinning of the stratigraphic segment and the evolution of more sandy facies in the direction of the Guyana Shield [9].

5.1 Petroleum geology

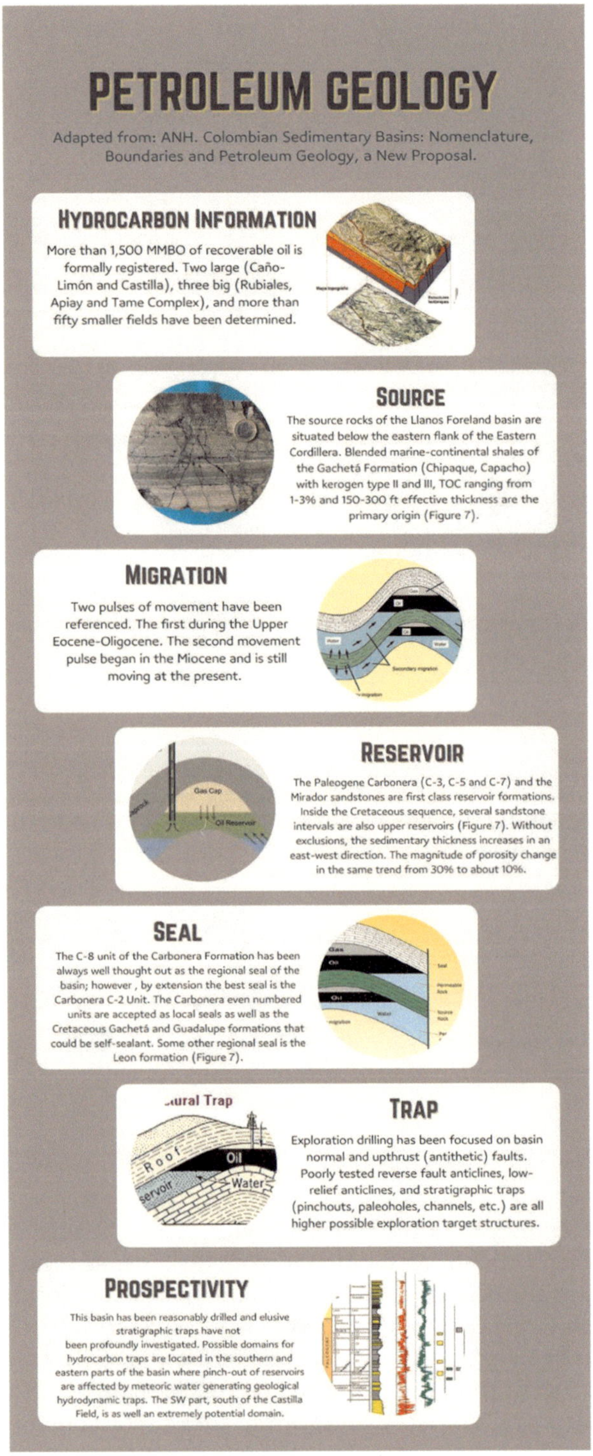

Figure 7.
Petroleum Geology—Eastern Basin [11].

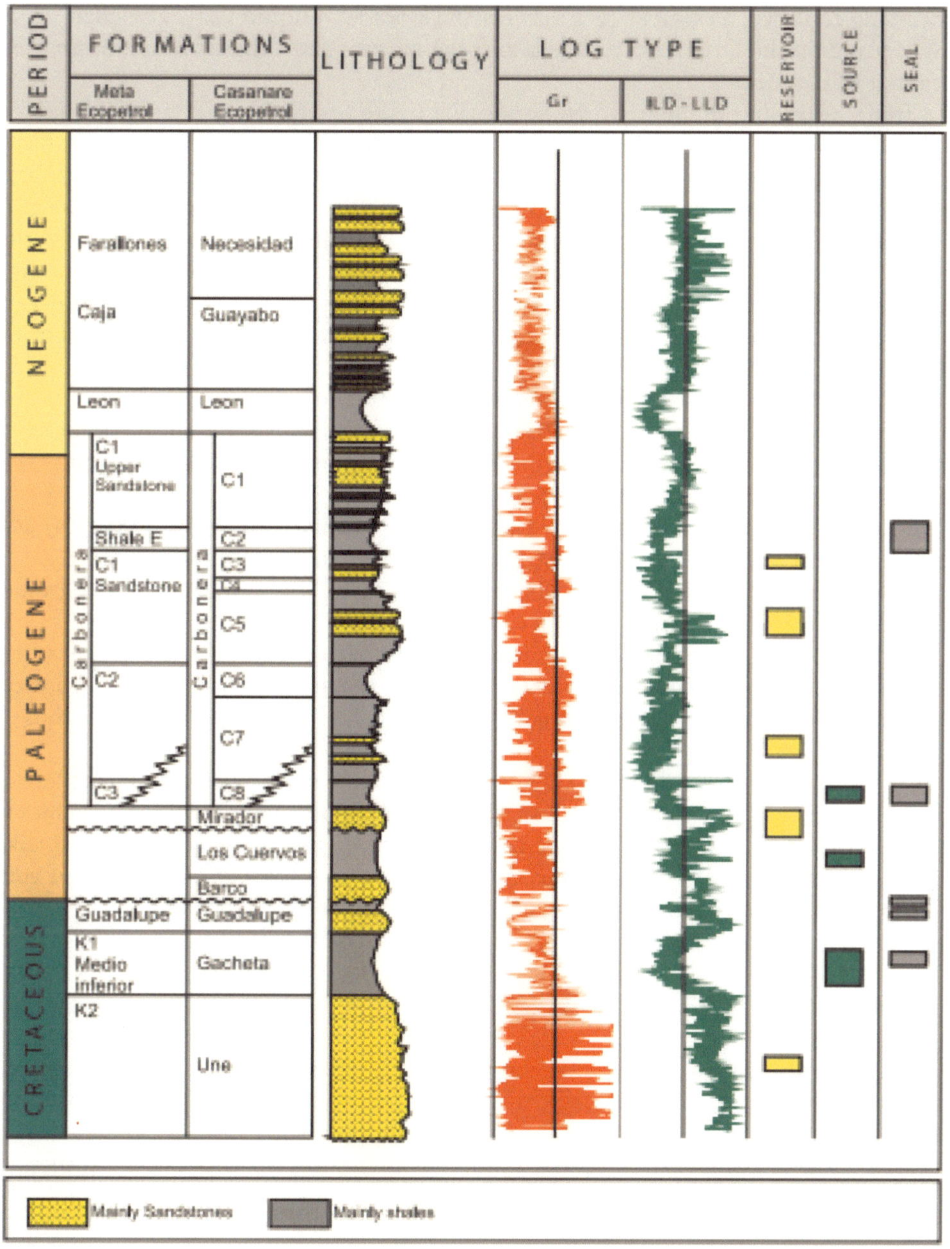

Figure 8.
Eastern Llanos Basin Petroleum system chart [11].

5.2 CCUS: EOR potential in the Eastern Llanos basin

The review of progress in the different projects of the hydrocarbons sector in the Eastern Llanos basin means that currently there is sufficient infrastructure to move forward the capture and injection of CO_2 for improved recovery, subtracting the construction of CO_2 liquefaction facilities with sufficient processing capacity, as well as facilities for the storage of liquid CO_2. The coal power plants in the Oriental Cordillera and the natural gas deposits in the basin are potential CO2 sources.

After researching the many formations that make up the Llanos Basin, it is concluded that the Carbonera formation is the most favorable for CCUS because it has sandy interbeddings that are not being used (levels C-7, C-5, C-3, and C-1) and that they can be favorable as a storage formation due to their storage capability in a future CO_2 capture project, all these are conditions to elaborated surveys of characterization of sand formation. Additionally, Carbonera is at a higher depth than the last potentially usable water source (Guayabo aquifer), isolated by clay layers that form natural hydraulic seals (levels C-2, C-4, C-6, and C-7) and covered by the regional seal of the León Formation (**Figure 6**).

The four clayey members of Carbonera add up to about 637 feet, which with low permeability provides the natural separation of the disposal area with the other units, guaranteeing the non-affectation to the "surface" aquifers in the Guayabo and Necesidad formations (**Figures** 7 and **8**) [12].

In the Carbonera formation, porosities are comparatively invariant (18–23%) as well as permeable (100–3500 mD), suggesting that the units that are projected as receptors have stable petrophysical characteristics that allow the entry of a volume of injected CO_2 without causing damage to the formation.

Porosities were measured through density and neutron curves taken in the wells. Regarding permeability, despite not having data in the aforementioned cores, a theoretical curve was obtained and then corrected and tuned with cores and logs from other blocks [13].

The Carbonera formation is not the only one with CO_2 storage expectations, there are also the Mirador, Barco, Guadalupe, and Une formations that have even better values for a storage formation, and therefore, the best hydrocarbon reservoirs found in the Llanos basin are in these formations; however, as CO_2 capture prospects, these formations could be used for secondary recovery (EOR/EGR) and obtain better results in production.

6. Conclusions

The geological depository of CO_2 should be considered as one of the most viable options to minimize the effects and emissions of polluting gases into the atmosphere related to fossil fuels.

At the time of reviewing the existing documentation on the method of capture and storage or sequestration of CO_2, it is clear that in Colombia there is no current regulation on CCUS–EOR technologies, given that they are still in the experimental phase and the projects that are starting are pilots, or there are very few who have come to implement the technique.

The maturity of the Llanos basin and its tectonic evolution, as well as the stratigraphic and petrophysical characteristics of the large oil and gas fields with hydrodynamic capture, and good regional seals, lead to the conclusion that in the basin there are formations with good possibilities to store or sequester CO_2; being the odd sandy intervals of the Carbonera formation and more specifically C-1 and C-7, the most suitable for the application of this method.

The technologies implemented within the hydrocarbons sector favor in a certain way the development of the CCUS method in Colombia, this is a considerable advantage compared to countries that do not have a presence of oil activity.

Author details

Jorge Eliecer Mariño Martínez[1*] and Luisa Epimenia Moreno Reyes[2]

1 Geological Engineering, Universidad Pedagógica y Tecnológica de Colombia, Boyacá, Colombia

2 Independent Geological Engineer, Tauramena, Casanare, Colombia

*Address all correspondence to: jorge.marino@uptc.edu.co; luisamorenor@gmail.com

References

[1] Rebeca T, Sundseth K, Bouman E, Czarnowska L, Mathisen A, Skagestad R, et al. Technical and environmental viability of a European CO_2 EOR system. International Journal of Greenhouse Gas Control. 2020;**92**:1–2

[2] IEA. Number of EOR PROJECTS in Operation Globally, 1971–2017. Paris: IEA;

[3] King Abdullah Petroleum Studies and Research Center. 2022. Available from: https://datasource.kapsarc.org/explore/dataset/large-scale-carbon-capture-projectsdatabase/. [Accessed: May 2, 2022]

[4] Advanced Resources International Inc. The U.S. CO_2 Enhanced Oil Recovery Survey. 2021. Available from: https://adv-res.com/pdf/ARI-2021-EOY-2020-CO2-EOR-Survey-OCT-21-2021.pdf

[5] Novak Mavar K, Gaurina-Međimurec N, Hrnčević L. Significance of enhanced oil recovery in carbon dioxide emission reduction. Sustainability. 2017;**13**:1800

[6] Bachu S. Sequestration of CO2 in geological media: Criteria and approach for site selection in response to climate change. Energy Conversion & Management. 2000;**41**:961–964

[7] Bachu S. Sequestration of CO2 in geological media in response to climate change: Road map for site selection using the transform of the geological space into the CO2 phase space. Energy Conversion and Management. 2022;**43**:89–90

[8] IEA. Can CO2-EOR Really Provide Carbon-negative Oil? Paris: IEA; 2019

[9] He R, Weizhong MA, Xinyu M, Liu Y. Modeling and optimizing for operation of CO2-EOR project based on machine learning methods and greedy algorithm. Energy Reports. 2021;7:3663–3664

[10] Advanced Resources International, Inc. Optimization of CO_2 Storage in CO_2 Enhanced Oil Recovery Projects. 2010.

[11] ANH. Colombian Sedimentary Basins: Nomenclature, Boundaries and Petroleum Geology, a New Proposal. Bogotá, Colombia: Colombia Mines and Energy Office; 2007

[12] Mendoza J, Bueno J. Modelado de causas geológicas generadoras de sobrepresión, aplicación para la formación Carbonera en el campo Cupiagua del Piedemonte Llanero Colombiano. Trabajo de grado Geología. Bucaramanga, Colombia. Universidad Industrial de Santander. Facultad de Ingeniería Fisicoquímica. Escuela de Geología; 2008

[13] Mariño-Martínez JE, Moreno-Reyes LE. Posibilidades de captura y almacenamiento geológico de CO2 (CCS) en Colombia – caso Tauramena (Casanare). Boletín de Geología. 2018; **40**(1):109-122

Chapter 2

Carbon Sequestration by Eucalypts in Florida, USA: Management Options Including Biochar and Associated Economics

Donald L. Rockwood, Kyle W. Fabbro, Martin F. Ellis, Monica Ozores-Hampton and Amir Varshovi

Abstract

Growth and economic models for *E. grandis* in mulch wood rotations, for *E. grandis* and *E. grandis* x *E. urophylla* cultivars grown as short-rotation woody crops (SRWC), including coppicing, for *E. grandis* in windbreaks (WB), with and without soil amendments including biochar (BC) and the slow-release fertilizer Green Edge (SRF, GE), and for *E. grandis* in dendroremediation applications estimated the above- and below-ground carbon sequestration potentials of these management options. The cultivars may sequester over 10 Mg of C/ha/year as SRWCs. Under assumed management costs and market conditions, SRWC management with BC is more profitable than operational culture if BC application costs are ≤$450/Mg. Longer rotations with less intensive management result in lower but still considerable sequestration and economic benefit. In WBs, *E. grandis* cultivars may sequester up to 34 Mg of C/ha in 3 years, with additional sequestration by amending soil with BC, GE, and BC + GE. Amending soil with BC derived from eucalypts is both a long-term sequestration strategy and an opportunity to increase plantation and crop productivity. Demand for sustainably produced BC is growing due to multiple applications beyond soil carbon sequestration.

Keywords: *Eucalyptus grandis*, *E. grandis* x *Eucalyptus urophylla* hybrid, mulch wood, short-rotation woody crops, carbon sequestration, management options, economic potential, biochar, slow-release fertilizer

1. Introduction

Eucalyptus species, the most widely planted hardwoods in the world [1], have considerable potential for sequestering carbon. For *E. urophylla* in Brazil and *E. globulus* in Spain, rotation length, number of coppice rotations, site quality, carbon credit, and discount rate influenced carbon sequestration value [2]. *E. urophylla* × *E. grandis* hybrids in subtropical China maximize sequestration in 12–15 year rotations [3]. In Pakistan, *E. camaldulensis* is one of the best sequestration options for marginal areas [4], and in northwest India, *E. tereticornis* used in agroforestry is a viable option for carbon mitigation [5]. In Portugal, sequestration by *E. globulus* plantations was smaller than that of their derived wood products [6].

In subtropical central and southern Florida, USA, (annual rainfall of ~1400 mm mainly during the summer, average maximum temperature of ~28°C, average minimum temperature of 18°C, and lowest temperature of −2°C), eucalypts have numerous potential applications. We previously described their potential for maximizing SRWC productivity through genetic improvement and site amendments, such as BC [7]. On former citrus and phosphate mined lands, *E. grandis* cultivars may have maximum mean annual increments (MAI_{max}) up to 78.2 green Mg/ha/year with an internal rate of returns (IRR) over 10% when grown as SRWCs [8].

BC improves many soil properties and thereby increases productivity [9–11], especially in sandy soils common to central and southern Florida [12, 13]. BC's numerous applications, including carbon sequestration [14], have considerable market potential.

Here, we expand our previous estimations of carbon sequestration by eucalypts with and without BC in Florida [15] by estimating (1) the economic potential for carbon sequestration by Eucalyptus planted in long-term mulch wood plantations, in more WBs, and in dendroremediation applications and (2) the responses to BC as a soil amendment with and without compost in additional field studies in Florida.

2. Materials and methods

Thirteen studies in central and southern Florida (27°–28°31′N, 80°–82°49′W) representing a range of *Eucalyptus* management options contributed to our analyses—(1) two *E. grandis* mulch wood studies, (2) two *E. grandis* cultivar planting density studies, (3) *E. grandis* x *E. urophylla* hybrid cultivar EH1 planting density demonstration, (4) EH1 fertilizer-planting density study at the Indian River Research and Education Center (IRREC), (5) five *E. grandis* WBs at Water Conserv II, Clermont, and the IRREC using BC, and (6) two *E. grandis* dendroremediation studies (**Table 1**).

2.1 *E. grandis* mulch wood plantations

In central and south Florida, *E. grandis* mulch wood plantations are typically established at moderate planting densities (1495–1794 trees/ha) with 7–10 year rotations and re-established after two or three coppice stages. Mulch wood plantation management intensity is low," with all cultural treatments, such as chemical site preparation, single-pass bedding, and N + P fertilization implemented prior to planting. Post-establishment silvicultural treatments, such as herbaceous chemical release and mid-rotation fertilization, are uncommon in most mulch wood plantations throughout the entire management cycle, including the coppice stages.

The carbon sequestration and yield potential of improved *E. grandis* open-pollinated (OP) family seedlings and cultivar G2 clones under low operational culture were based on field demonstration Studies 1A and 1B, respectively, established on bedded cutover flatwoods sites on poorly drained, sandy Spodic soils. The planting density was 1495 trees/ha at a tree spacing of ≈ 1.8 m within row × 3.7 m between beds. Stem wood green weight estimates were based on felled tree samples and stand-level, whole-stem green weight estimates were fitted to the equation below using nonlinear regression [8]:

$$B(t) = e^{\left[b + c\times\ln(t) - d\times t\right]} \tag{1}$$

Study	Location	Option	Genotype	Soil	Culture	Density	Age
1A	Palmdale	MW	Seedlings	Sandy	B, H, F	1495	8 yrs
1B	Palmdale	MW	Cultivars	Sandy	B, H, F	1495	7 yrs
2A	Ft Meade	SRWC	Cultivars	Clay	B, F, H, I	2148, 2872, 4305	48
2B	Indiantown	SRWC	Cultivars	Sandy	B, H	1436, 2148, 2872, 4305	48
3	Hobe Sound	SRWC	Cultivars	Sandy	B, F, H, I	1181, 2471	81
4	Ft Pierce	SRWC	Cultivars	Sandy	F, H	1196, 1794, 3588	47, 28
5A	Winter Garden	WB	Cultivars	Sandy	F, I	~2778	52
5B	Winter Garden	WB	Cultivars	Sandy	F, I	~2778	16
5C	Winter Garden	WB	Cultivars	Sandy	F, I	~2778	16
5D	Clermont	WB	Cultivars	Sandy	F, H, I	~3472	74
5E	Ft Pierce	WB	Cultivars	Sandy	F, BC, H, I	~4630	37
6A	Tampa	DR	Cultivars	Sandy	H, I	2778	44
6B	Belle Glade	DR	Cultivars	Muck	H, I	4444	12

Table 1.
Description of 13 Eucalyptus studies in Florida: location in FL, management option (mulch wood = MW, dendroremediation = DR). genotypes involved, soil type, culture (B = bedded, F = fertilized, H = herbicided, I = irrigated), planting density (trees/ha), and age (months) at final measurement.

where $B(t)$ = whole-stem green weight (metric tons/ha), t = stand age (years), and b, c, d are estimated parameters.

Stem wood carbon content was estimated as 25% of stem green weight. On sandy soils, 78% of total C sequestration for *E. grandis* was assumed to be in stem wood [16]. **Table 2** outlines the operational silvicultural treatments previously described, their associated costs, and stumpage and carbon price assumptions. Three coppice stages were assumed with coppice yields projected to be 80, 60, and 40% of the original stand for stages 2, 3, and 4, respectively.

2.2 *E. grandis* cultivar planting density studies

Studies 2A and 2B on a phosphate mine clay settling area and former citrus beds, respectively, assessed the effect of planting densities (**Table 1**) on the biomass production of three *E. grandis* cultivars (G2, G3, and G5). Stand-level whole-stem green weight estimates (based on felled and standing trees in Florida) for each planting density were calculated by Eq. (1) using nonlinear regression [8]. The carbon content of stem wood was again assumed to be 25% of stem green weight.

The economic assumptions in **Table 2** were coupled with the assumptions that 78% of total C sequestration for *E. grandis* on sandy soils is in stem wood [16] and that response to BC followed that observed in Study 5E. Yields in two coppice rotations

Activity	Management option	
	Mulch wood	SRWC
Land preparation (start-up cost)	$618/ha	$1236/ha
Chemical site prep (beginning of each cycle)	$173/ha	$297/ha
Weed control (beginning of coppice stage)		$136/ha
Planting cost	$0.08/tree	$0.08/tree
Seedlings	$0.30/tree	
Clones	$0.70/tree	$0.70/tree
Fertilization (beginning of each cycle)	$223/ha	$223/ha
BC application (one-time start-up cost—low)		$750/Mg
BC application (one-time start-up cost—high)		$1000/Mg
Stumpage price	$13/green Mg	$13/green Mg
Carbon credit [17]	$5/Mg C	$5/Mg C

Table 2.
Management costs and timber stumpage and carbon credit assumptions for two management options for E. grandis grown on sandy soils in central and southern Florida.

were projected to be 80 and 60% of the original stand for fertilization only and 90 and 80% of the original for fertilization + BC [18]. The application of BC priced at $750 and $1,000/ton assumed a 7% growth increase per ton of BC.

2.3 EH1 planting density study

Study 3, an intensively managed 8+ ha demonstration planted in May 2011 on sandy former citrus beds at two planting densities (**Table 1**) was monitored through December 2017. Stand-level whole-stem green weight equations [19] used periodic data through 81 months to model growth scenarios at the original two planting densities and an intermediate density of 1,794 trees/ha assuming original and two coppice rotations for each density, with the two coppices growing at 90% and 80% of the original planting. EH1 stem wood carbon content was assumed to be 25% of stem green weight, and 78% of total carbon sequestration was in stem wood.

2.4 EH1 fertilizer-planting density-coppicing study

EH1, planted in June 2015 on a sandy former pasture in five 3-row (26 trees/row) plots receiving one of five fertilizers (control, GE 6-4-0 + micronutrients at 112, 224, and 336 kg of N/ha rates, and diammonium phosphate equivalent to 336 kg of N/ha) and two replications of 5-tree row plots of three planting densities (1196, 1794, and 3588 trees/ha), was coppiced in June 2019. The interior row of each plot was periodically measured for tree size, and number of coppice stems/stool at least half the DBH of the largest stem, through November 2021.

Given eucalypt's high productivity and their use for traditional forest products and because economic feasibility is one of several conditions for a sustainable BC system [20], our financial analysis goal using Land Expectation Value (LEV) and IRR in Sections 2.1–2.4 was to estimate the cost of potential carbon sequestration by *Eucalyptus* genotypes with and without BC as a soil amendment.

2.5 *E. grandis* WBs

Two-row WB 5A, consisting of four *E. grandis* cultivars in 20-tree plots (two staggered rows 2.4 m apart with 10 trees at 1.5 m spacing within rows) systematically positioned in 14 replications, was established in June 2009 at Water Conserv II. All replications were irrigated with reclaimed water. The cultivars were measured periodically through 52 months for height and DBH. Assumed sequestration in roots was ≈10.3% of total aboveground sequestration [16].

In June 2012, two-row WBs (5B and 5C) composed of four *E. grandis* cultivars (G1, G2, G3, and G4) in one row and up to eight *Corymbia torelliana* progenies in an adjacent staggered row 2.4 m away were established around two Water Conserv II Rapid Infiltration Basins (RIB 2-3 and RIB 3-2). The trees were subsequently irrigated with reclaimed water. From the 290 clones of the cultivars replicated up to five times in row plots around RIB 2-3 and from the 308 clones replicated up to five times in row plots around RIB 3-2, typically 10-tree subsets in the row plots were measured periodically.

On March 30, 2014, two-row WB 5D was established at a citrus grove following Roundup application in mid-March. At 2.4 m spacing, 68 G3s were planted in the interior (north) row and 68 *C. torelliana* in the staggered (1.2 m offset) exterior (south) row. The trees were subsequently irrigated for 4 years and measured in May 2020.

Two-row WB 5E, consisting of three *E. grandis* cultivars in one row and four *C. torelliana* progenies in an adjacent row offset 1.2 m away, was established in July 2017 to assess BC and GE as silvicultural management options. Initially a randomized complete block design with four complete and one incomplete replications of the cultivars at 1.8 m within row spacing, in February 2018, all four complete replications received GE (6-4-0 + micronutrients equivalent to 336 kg of N/ha) and two interior replications also received 11.2 Mg/ha of GCS' Polchar BC by rotovating the two treatments into the soil to a 20 cm depth between and within 1.2 m of the two rows. The incomplete replication served as a control. The cultivars were measured periodically through June 2020.

2.6 *E. grandis* dendroremediation studies

Two dendroremediation studies (**Table 1**) represent the potential use of *Eucalyptus* for managing wastewater. Study 6A had 44-month-old *E. grandis* cultivars G2 and G3 at 2.4 × 1.5 m in sandy soil in a stormwater retention pond in Tampa, FL, at the Tampa Port Authority (TPA). Study 6B on muck soil at the Everglades Research and Education Center (EREC) at Belle Glade, FL, included two *E. grandis* cultivars (G3 and G4) planted at a 1.5 × 1.5 m inside an agricultural runoff collection pond and measured for tree size and survival at 12 months. Above- and below-ground carbon sequestration was estimated as described in Section 2.2.

2.7 Other BC field studies

Seven recent BC studies, all on sandy soil, are described in **Table 3**. GCS' Polchar BC was used for studies 7A, 7B, 7C, 7D, and 7E. Four studies (7A, 7C, 7D, 7F) involved levels of BC only, two (7B, 7E) also had GE alone and in combination with BC, and one (7G) included BC/compost mixes. The crops and soils were monitored periodically for up to two years.

Study 7E (**Table 3**) had two replications of four treatments: 0, GE equivalent to 336 kg of N/ha, 11.2 Mg/ha of GCS' Polchar BC, and GE + BC. The BC was banded

Study	Location	Amendments	Crop	Soil	Culture
7A	Gainesville	0, 11.2 mt/ha BC, 11.2 mt/ha BC twice	Vegetables	Sandy	Open field
7B	Gainesville	0, 11.2 mt/ha BC, GE, 11.2 mt/ha BC + GE	Perennial peanut	Sandy	Open field
7C	Old Town	0, 5.6, 11.2, 16.8, 22.4 mt/ha BC	Sorghum	Sandy	Open field
7D	Old Town	0, 11.2 mt/ha BC	Bahiagrass	Sandy	Open field
7E	Gainesville	0, 11.2 mt/ha BC twice, GE, 11.2 mt/ha BC twice BC + GE	Slash pine, Cypress	Sandy	Bedded
7F	Immokalee, Myakka City	0 and 286 kg/ha BC	Tomatoes	Sandy	Plasticulture/ Open field
7G	Immokalee	0, 446, and 892 kg/ha BC, BC at 2.5 and 5% plus compost at 4.5 Mg/ha	Citrus	Sandy	Open field

Table 3.
Description of field studies receiving BC, GE, and/or compost—location in FL, amendments, crop, soil type, and culture.

and incorporated into beds twice, and the GE was banded on top of fully formed beds. Soil samples were taken in January 2021 after all treatments has been applied.

The five experiments in Study 7F (**Table 3**) were conducted in two major commercial tomato production areas during the fall and winter of 2018–2019. Plastic beds (20 and 18 cm high in the middle and on the edges, respectively, and 81 cm wide) were formed at 1.8 m centers. Following formation, they were fertilized with a fertilizer/BC mixture (BC from coconut shells blended with the fertilizer at the blending facility at 268 lbs/ha), fumigated with 1,3-Dichloropropene and Chloropicrin (40:60) at a rate of 123 and 134 kg ha^{-1}, and covered with virtually impermeable film. In all trials, pre-plant dry fertilizer (ammonium nitrate, triple superphosphate, and potassium sulfate plus micronutrients) was broadcast as "bottom mix" and two fertilizer bands were applied on the bed shoulders as "top mix" for a total nitrogen-phosphorus-potassium (N-P-K) of 207-49-344 kg ha^{-1}. Fertigation supplemented the pre-plant fertilizer with 112-0-167 kg ha^{-1} N-P-K from tomato flowering to the first harvest. Roma-type tomatoes were harvested two to three times at the mature-green stage and graded into marketable sizes and weighed separately according to USDA specifications: extra-large (>7.00 cm), large (6.35–7.06 cm), and medium (5.72–6.43 cm).

Study 7G's three BC levels and two compost/BC mixes (**Table 3**) were applied annually to "Valencia" bud-grafted to "US812" planted in spring 2016. Tree growth measurements consisted of trunk diameter and fruit yield. Fruit mass per plot was assessed annually by weighing harvested fruit from entire plots using a Gator Deck scale (Scale Systems, Novi, MI).

3. Results

3.1 *E. grandis* mulch wood plantations

The MAI_{max} and biological rotation age for OP seedlings and G2 clones were 10.5 green Mg/ha/year at age 8.0 years and 16.5 green Mg/ha/year at age 7.0 years, respectively, and their associated total carbon sequestrations at MAI_{max} were 27.0 and 37.0 Mg C/ha (**Figure 1**). The observed yields corresponded to site index (base age 8 years) values of 15.2 and 21.3 m for the seedlings and clones, respectively. LEVs at an 8% real discount rate, with and without carbon, ranged between −$731/ha and −$517/ha with IRRs between 4.3 and 5.9% (**Table 4**).

These yields for improved *E. grandis* OP seedlings were similar to earlier *E. grandis* spacing trial results in south Florida [21]. Under operational culture and without carbon credits, stumpage prices ≥ $15/green Mg would favor clonal deployment over family forestry with IRRs exceeding 6.1%. Clonal deployment could generate higher LEVs at stumpage prices as low as $13/green Mg with carbon credits included. Family forestry under operational culture and without carbon credits is favorable when stumpage prices are <$15/green Mg and can exceed a 6% IRR when stumpage prices are ≥$16.30/green Mg.

3.2 *E. grandis* cultivar planting density studies

On former citrus lands and phosphate mined clay settling areas in central and south Florida, *E. grandis* cultivars had MAI_{max}s as high as 78.2 green Mg/ha/year with

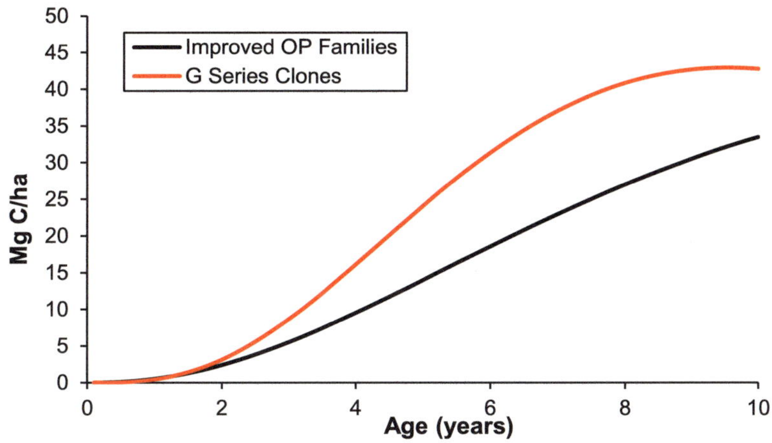

Figure 1.
Estimated total (stem + crown + roots) carbon sequestration (C, Mg/ha) for mulch wood plantations of E. grandis OP families and G2 clones established at 1,495 trees/ha on poorly drained, sandy Flatwoods sites in South Florida.

Genotype	OP families	G2 cultivar
Total carbon sequestration (C, Mg/ha)	27.0	37.0
MAI_{max} (green Mg/ha/yr)—rotation age (yrs)	10.5–8.0	16.5–7.0
LEV ($/ha)—IRR (%) without carbon credit	−$652/ha—4.3	−$731/ha—4.9%
LEV ($/ha)—IRR (%) with carbon credit	−$519/ha—5.1%	−$517/ha—5.9%

Table 4.
Estimated total carbon sequestration at MAI_{max}, MAI_{max}, and associated rotation age, and LEVs at 8% real discount rate and associated IRRs with and without carbon credits, for mulch wood plantations of E. grandis genotypes OP families and G2 cultivar established at 1495 trees/ha on bedded flatwoods soils.

associated IRRs greater than 10% [8]. Total carbon sequestration estimates ranged from 38 to 95 Mg/ha at the time of MAI_{max}, with longer-term totals over 100 Mg/ha in 6 years, depending on cultivar, site, planting density, and harvest age.

The effects of adding BC as a soil amendment on sandy soils and of applying carbon credits were assessed (**Table 5, Figure 2**). Because BC increased growth and decreased time of MAI_{max}, estimated cumulative carbon sequestration with BC decreased as rotation length decreased; for example, at 2148 trees/ha, sequestration was 69.4 Mg/ha C in 4.9 years without BC and 61.9 Mg/ha C in 3.5 years with BC. Under current market conditions in central and southern Florida, intensive management with BC will be more profitable than operational culture if BC application costs are ≤$450/Mg. If BC costs $450/Mg, for example, then the LEV for 4305 trees/ha with BC will exceed the LEV of 2148 trees/ha under operational culture. Increased stumpage prices and carbon credits and/or lower silvicultural management costs favor an intensive BC regime under current application costs.

Increased stumpage price and low BC cost ($750/Mg) favor a higher planting density under intensive management over the current mulch wood/moderate planting densities under operational culture. For example, a planting density of 4305 trees/

Response—associated response	**Planting density (trees/ha)**			
	1436	**2148**	**2872**	**4305**
Fertilization only				
Total carbon sequestration (C, Mg/ha)	52.7	69.4	73.2	77.4
MAI_{max} (green Mg/ha/yr) —rotation age (yrs)	39.0–4.3	45.1–4.9	51.8–4.5	63.2–3.9
LEV ($/ha)—IRR (%) without carbon credit	282–8.8	413–8.9	216–8.4	−712 to 6.7
LEV ($/ha)—IRR (%) with carbon credit	848–10.3	1054–10.3	963–9.9	216–8.4
Fertilization + 6.2 Mg/ha of BC				
Total carbon sequestration (C, Mg/ha)	55.5	61.9	69.5	92.6
MAI_{max} (green Mg/ha/yr)—rotation age (yrs)	55.2–3.2	56.3–3.5	73.8–3.0	92.1–3.2
Fertilization + 6.2 Mg/ha of BC @$750/Mg				
LEV ($/ha)—IRR (%) without carbon credit	−2217 to 5.6	−2803 to 5.2	−1914 to 6.2	−1464 to 6.7
LEV ($/ha)—IRR (%) with carbon credit	−1306 to 6.6	−1885 to 6.1	−690 to 7.3	55–8.0
Fertilization + 6.2 Mg/ha of BC @$1,000/Mg				
LEV ($/ha)—IRR (%) without carbon credit	−3761 to 4.6	−4348 to 4.3	−3458 to 5.2	−3008 to 5.7
LEV ($/ha)—IRR (%) with carbon credit	−2850 to 5.5	−3429 to 5.2	−2234 to 6.2	−1489 to 6.9

Table 5.
Estimated total (stem + crown + roots) carbon sequestration at MAI_{max}, MAI_{max}, and associated rotation age, and LEV and associated IRR for E. grandis cultivars at two cultural intensities (fertilization and fertilization + BC), with and without carbon credits ($5/Mg C), two BC prices ($750 and 1000/Mg), and four planting densities on sandy soils in central and southern Florida.

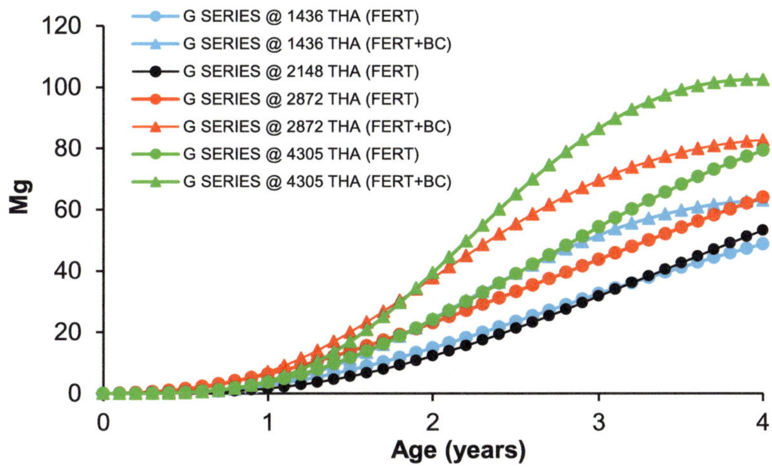

Figure 2.
Estimated total (stem + crown + roots) carbon sequestration (C, Mg/ha) for G Series E. grandis cultivars for 4 years under four planting densities and two cultural regimes (fertilization only vs. fertilizer + BC) on sandy bedded former citrus lands in central and southern Florida.

ha under an intensively managed BC regime can be more profitable (LEV = $3357/ha) than the moderate 2148 trees/ha planting density under operational culture (LEV = $2459/ha), assuming the $18/green Mg stumpage price observed in central and southern FL mulch wood markets (no carbon credits), BC application cost of $750/Mg, and 8% real discount rate (and the same management costs outlined in **Table 2**).

3.3 EH1 planting density study

Through 81 months, the higher 2471 tree/ha density increased the yield of intensively managed EH1 [7]. Maximum annual biomass yields and time to those maxima were directly and inversely, respectively, related to planting density: >58 green Mg/ha/year in 3.7 years at 2471 trees/ha vs. 44 at 5.0 years for 1181 trees/ha. Associated total carbon sequestration estimates followed somewhat similar trends: 77.2 Mg/ha C at 4.7 years for 2471 trees/ha vs. 75.8 Mg/ha at 5.5 years for 1181 trees/ha (**Table 6**, **Figure 3**). Assessing the economic feasibility of EH1 SRWCs at a stumpage price of $13/Mg and without BC, LEVs, and IRRs increased with carbon credit and were highest at an intermediate planting density.

3.4 EH1 fertilizer-planting density-coppicing study

Planting density consistently influenced tree size, and the highest planting density had the smallest tree DBH at the 47-month harvest of the original rotation ([7], **Table 7**). However, carbon sequestration at 47 months was greatest at the 3588 density.

While planting density usually did not influence coppice stem DBH and number, at 23 months, the DBHs of the largest coppice stem/stool (**Table 7**) were similar to tree DBH at the same age in the original rotation. Should that trend continue and the number of coppice stems/stool with DBH at least half that of the largest stem exceeds one, coppice carbon sequestration at each planting density would surpass that of the original rotation.

Response—associated response	Planting density (trees/ha)		
	1181	1794	2471
Total carbon sequestration (C, Mg/ha)	75.8	76.3	77.2
MAI_{max} (green Mg/ha/yr)—rotation age (yrs)	47.1–5.5	52.1–5.0	56.0–4.7
LEV ($/ha)—IRR (%) without carbon credit	2292–13.5	2913–15.0	1871–11.7
LEV ($/ha)—IRR (%) with carbon credit	2959–14.9	3665–16.6	2687–13.2

Table 6.
Estimated total carbon sequestration at MAI_{max}, MAI_{max} and associated rotation age, and LEV at 8% real discount rate and associated IRR with and without carbon credits ($5/Mg C) for EH1 under operational culture without BC and three planting densities on sandy bedded former citrus lands in southern Florida.

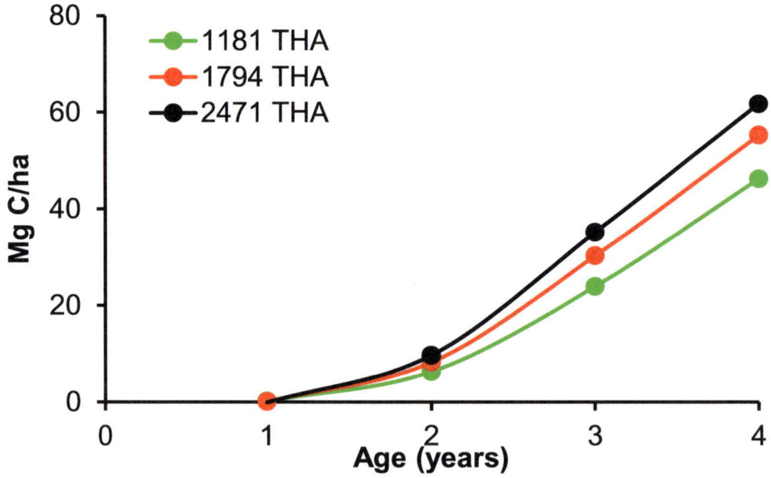

Figure 3.
Estimated total (stem + crown + roots) carbon sequestration (C, Mg/ha) for EH1 at three planting densities (trees/ha, THA) through 4 years in study 4 without BC.

Response	Planting density (trees/ha)		
	1196	1794	3588
47 months after planting			
DBH (cm)	15.4	13.5	11.4
Total carbon sequestration (Mg/ha C)	50.9	53.9	68.7
23-month-old Coppice			
DBH (cm)	8.3	8.4	6.4
No. of stems	4.1	3.8	2.7
Total carbon sequestration (Mg/ha C)	16.9	24.8	18.5

Table 7.
DBH and estimated total carbon sequestration of 47-month-old original and DBH and number of coppice stems of E. urophylla x E. grandis cultivar EH1 in Study 4.

3.5 *E. grandis* in WBs

WBs 5A, 5B, 5C, and 5D were measured from as young as 4 months to as old as 74 months (**Table 8**). Because the four *E. grandis* cultivars in WB 5A had similar sizes at each measurement age, their carbon sequestration estimates were averaged for each age. Sequestrations increased with age, reaching 12 Mg/ha C at 52 months. In WB 5B, because the cultivars were bigger in RIB 2–3, at 16 months, the cultivars had higher sequestration in RIB 2–3; both 16-month sequestration levels approximated the 18-month level in WB 5A. In WB 5C at age 74 months, cultivar G3 grew well and sequestered 33 Mg/ha C in just over 6 years.

Sequestration estimates in these three WBs were influenced by the planting density presumed for the three WBs. While the within-row spacing and distance between rows were known for each WB, the area occupied by each WB tree was speculative and was set to 652 trees/ha for each WB. Had a higher planting density been used, the sequestration estimates would be higher.

Soil amendments in WB 5E caused large early soil nutrient, tree nutrient, and tree growth responses by three *E. grandis* cultivars [7], with sequestration of up to 34 Mg/ha of C in 37 months with GE + BC (**Table 9**). GE and especially BC + GE greatly enhanced the nutrient properties of this inherently poor sandy soil.

GE greatly increased tree DBH and total carbon sequestration compared to the control, and GE + BC further increased DBH by 3.3 cm and C by 14 Mg/ha, respectively. Carbon sequestration from GE is primarily above ground while carbon sequestration by GE + BC is both above ground and in the soil. Assuming that all the BC applied remained in the soil, GE + BC increased total carbon sequestration by nearly 33% to some 45 Mg/ha of C.

Age (months)	Height (m)	DBH (cm)	Carbon sequestration (Mg/ha C)		
			Above ground	Below ground	Total
WB 5A: two *E. grandis* rows					
18	5.5	5.9	.69	.07	.76
25	7.4	7.2	1.53	.16	1.69
52	14.3	13.4	10.86	1.14	12.00
WB 5B—RIB 2-3: one *E. grandis* row, one *C. torelliana* row					
4	1.1				
8	2.0	1.0			
16	6.0	8.2	.83	.09	.92
WB 5C—RIB 3-2: one *E. grandis* Row, one *C. torelliana* row					
4	1.2				
8	1.8	.8			
16	4.9	7.2	.51	.05	.56
WB 5D: one *E. grandis* row, one *C. torelliana* row					
74	24.1	24.2	30.1	3.0	33.1

Table 8.
Tree height and DBH and estimated carbon sequestration at various ages of E. grandis cultivars in four WB studies.

Response—associated response	Treatment		
	Control	GE	GE + BC
DBH (cm)	5.8	10.3	13.6
Total carbon sequestration (C, Mg/ha)	2.8	19.6	33.6
MAI_{max} (green Mg/ha/year)—rotation age (years)	3.4–2.7	17.3–3.7	32.5–3.3

Table 9.
Tree DBH, estimated total carbon sequestration at 37 months, and MAI_{max} and associated rotation age of E. grandis cultivars receiving Control, GE, GE + BC treatments in WB Study 5E.

3.6 *E. grandis* dendroremediation studies

Studies 6A and 6B provided 12- and 44-month sequestration estimates, respectively (**Table 10**), for very different soil types and planting densities. Sequestration in 6A was 12 Mg/ha C at 44 months, or 12 Mg/ha C annually, on a sandy retention pond at 2778 trees/ha, while in 6B it was 12 Mg/ha C at 12 months on muck soil at 4444 trees/ha.

3.7 Other BC field studies

Five recent amendment studies involving BC, GE, and/or compost are summarized in **Table 11**. As suggested by Study 7A, notable soil and plant responses to BC may take up to 2 years, although BC immediately increased soil organic matter in Studies 7B, D, and E. Studies 7C and 7B had varied responses to BC rates.

In Study 7F, BC at 286 kg/ha only impacted the marketable yields in one out of five tomato trials (**Table 12**). Blending the BC with the broadcasting fertilizer application reduced the expense of an extra passing applying the BC; however, the rates were too low to produce an increase in marketable tomato yields. Similar studies indicate the use of BC was an effective and productive soil amendment as compared to compost [23–27]. Future trials with higher BC rates may impact tomato yields positively as may continue with yearly BC application at a lower rate.

Study 7G's first-year data indicated no differences in plant growth, but 892 kg/ha BC produced the highest fruit yields (**Table 13**), as application rates in this trial were too low to have a significant yield impact in the first year. Compost application in sandy

Cultivar	Height (m)	DBH (cm)	Survival (%)	Carbon sequestration (Mg/ha C)		
				Above	Below	Total
6A: 44-month-old at 2778 trees/ha [22]						
G2	9.6	6.7	67	7.2	0.7	7.9
G3	9.3	8.0	100	15.1	1.5	16.6
6B: 12-month-old at 4444 trees/ha						
G3	6.1	4.9	100	5.3	0.5	5.8
G4	5.7	4.2	100	3.3	0.3	3.6

Table 10.
Tree height, DBH, survival, and estimated above- and below-ground and total carbon sequestration of E. grandis cultivars in two dendroremediation studies.

Response	BC Level (mt/ha)					GE	BC + GE
	0	5.6	11.2	16.8	22.4		
7A—cauliflower (22 months after first BC application)							
Soil NO3-N (ppm)	2.45		3.44		2.19		
Leaf N (%)	4.68		5.29		4.70		
7B—perennial peanut (21 months after application)							
CEC (meq/100 g)	7.9		6.8			7.4	9.2
Soil OM (%)	1.42		1.93				
7C—sorghum (4 months after application)							
Soil NO3-N (kg/ha)	1.47	3.69	2.16	2.05	2.25		
Soil Ca (kg/ha)	3015	3094	3670	3255	3525		
Soil CEC (meq/100 g)	7.6	8.0	9.0	8.2	8.8		
7D—bahiagrass (13 months after application)							
Soil K (kg/ha)	21		166				
Soil OM (%)	0.8		1.3				
Soil CEC (meq/100 g)	5.0		9.1				
7E—slash pine/cypress (after application)							
Soil OM (%)	.60		1.06			.67	2.26
Soil CEC (meq/100 g)	3.1		4.4			3.3	6.7

Table 11.
Soil and plant responses in five BC and/or GE studies in Florida.

Tomato type	Season	Number of harvests	Yield response
Roma	Fall	2	Increase
Roma	Fall	2	No differences
Round	Fall	2	No differences
Round	Winter	3	No differences
Round	Fall	3	No differences

Table 12.
Effect of BC on the marketable yields of Roma and round-type tomatoes.

Soil Amendment		Trunk diameter (cm)	Fruit yield (kg/ha)
BC Level (kg/ha or %)	Compost (%)		
0	0	15.6	1600.0
446	0	15.1	2057.1
892	0	14.0	3200.0
2.5%	97.5	13.2	2514.3
5%	95.0	15.4	2057.1

Table 13.
First-year trunk diameter and fruit yield of Valencia/US812 in response to five BC/compost soil amendments.

soils had a positive impact on soil and crops elsewhere in Florida [28–32]. Long-term compost application at higher rates will promote soil health and increase yield [33, 34].

4. Discussion

The above- and below-ground carbon sequestration of productive eucalypts worldwide depends on site conditions and management options, such as genotype, cultural intensity, planting density, and rotation length (**Table 14**). Several types of *Eucalyptus* have promise as SRWCs in Florida [39, 40], including cultivars, such as *E. grandis* G3 and *E. grandis* x *E. urophylla* EH1. EH1 on former citrus beds and managed at relatively low intensity, for example, could sequester over 20 Mg of C/ha/year. The Florida WB and dendroremediation estimates are influenced by their assumed planting densities. Plantations, though, have well-defined planting densities that offer more reliable carbon sequestration values. As the other Florida examples demonstrate, sequestration estimates vary due to tree age, size, management, and genotype. Longer first and coppice rotations may maximize sequestration [3].

Our carbon sequestration estimates for *E. grandis* and *E. grandis* x *E. urophylla* in Florida approximated their potential, as several assumptions were involved. Green weights for *E. grandis* x *E. urophylla* were derived from Florida field data by a species-specific equation from Swaziland [19]. Stem wood carbon content was an assumed percentage of green weight. Above- and below-ground sequestration proportions

Species	Location	Management	Age	BC	C
E. grandis	Palmdale, FL	Seedlings, MW	8.0	No	27.0
E. grandis	Palmdale, FL	G2, MW	7.0	No	28.0
E. grandis	Ft Meade, FL	G3, SRWC	6.0	No	112.8
E. grandis	Indiantown, FL	G3, SRWC	3.9	No	77.3
E. grandis x urophylla	Hobe Sound, FL	EH1, SRWC	4.0	No	81.9
E. grandis x urophylla	Ft Pierce, FL	EH1, SRWC	3.9	No	68.7
E. grandis	Ft Pierce, FL	G3, double row WB	3.1	Yes	33.6
E. grandis	Winter Garden, FL	G3, double row WB	4.3	No	11.3
E. grandis	Winter Garden, FL	G3, double row WB	1.3	No	0.83
E. grandis	Winter Garden, FL	G3, double row WB	1.3	No	0.514
E. grandis	Clermont, FL	G3, double row WB	6.2	No	33.1
E. grandis	Tampa, FL	G3, DR	3.7	No	15.1
E. grandis	Belle Glade, FL	G3, DR	1.0	No	5.5
E. grandis	South Africa [35]		10	No	47
E. grandis	South Africa [35]		25	No	270
E. spp	Southern China [36]	Various	Var.	No	100
E. grandis x urophylla	Southern China [37]		6–8	No	>70
E. tereticornis	India [38]		4	No	116

Table 14.
Comparison of estimated above-ground carbon sequestration (C, Mg/ha) by Eucalyptus species in Florida with and without BC to sequestration elsewhere under varied managements and ages (years).

were based on *E. grandis* in Brazil [16]. Below-ground sequestration estimates assumed no soil C flux. Similar assumptions were used for sequestration estimates in South Africa [35] and China [36, 37].

In combination with the carbon sequestered in trees, cost estimates of sequestration in *Eucalyptus* plantations by using wood BC as a soil amendment were previously estimated at ~$5/Mg of BC added per ha [7]. Using the intensively managed *E. grandis* plantation with 4305 trees/ha (**Table 4**), a single planting cycle, and three coppices, the estimated cost for using wood BC at $750/ton as a soil amendment to accelerate sequestration is ~$4/Mg of C sequestered. If a second planting cycle is included, the with and without BC cost comparisons are very similar. In a scenario with a minimum of two planting cycles and BC less than $650/Mg, there is an economic incentive to use BC as a soil amendment to accelerate and increase carbon sequestration. These costs are less than the $30–50/ton estimated in 2005 for US forestry sequestering up to 500 million tons of C/year [41]. In 2015, the California Air Resources Board listed C sequestration credits at $12–13/ton [42].

Converting woody biomass into long-term forest products, such as BC, can be a critical component of carbon sequestration. BC produced from hardwoods has a soil residence time exceeding 1000 years [43]. In South Africa, carbon sequestration by *Eucalyptus* and their long-lived forest products may equally result in offsetting some 2% of the country's carbon emissions [35].

Because BC quality influences BC impact on soil properties and plant productivity, Study 5E used GCS' premium BC, which was produced from roundwood, was highly porous, and had high carbon content (93–95% fixed carbon on a dry weight (DW) basis), low ash content (2–3% DW), and high surface area (585–630 m^2/g).

BC enhances the nutrient properties of Florida's sandy soils as well as the nutrient status of *E. grandis*, especially when applied together with organic amendments, such as GE and/or chemical fertilizers. However, because soil C may decrease as *Eucalyptus* plantations mature [35], BC incorporation into plantation soil can be beneficial. BC application to the soil in Poland is viewed as an important component of the region's circular economy and means of counteracting climate change [44].

The relatively low levels of BC in Studies 7F and 7G had minimal impact on yield. Because both compost and BC improve soil physical properties (water-holding capacity, soil structure, and bulk density), soil chemical properties (cation exchange capacity and plant nutrient availability), and soil biological properties (microbial activity), they could, at higher levels, potentially mitigate symptoms of citrus greening, such as asymmetrical chlorosis of the leaves, foliar micronutrient deficiencies, root degeneration, leaf, and fruit drop and eventually dieback and sometimes death [45].

BC has benefited many crops. BC produced from *E. camaldulensis* increased critical soil properties and groundnut yield in Senegal [46]. BC applications have increased the yields of corn [47, 48], safflower [49], rice [50], cypress [51], and rubber [52]. BC-blended compost significantly improved crop quantity and quality in Europe [53]. In Florida, oak-derived BC as a soil amendment combined with standard fertilizers enhanced lettuce (*Lactuca sativa*) productivity in a greenhouse study [7], and Studies 7A-7E suggest that plant and soil nutrients may be enhanced by GE, BC, and/or BC + GE applications.

The SRF GE has also been used in several specialty crops, such as turfgrass, citrus, and landscape plants. Environmental concerns regarding quick release (soluble) fertilizers will continue to increase demand for SRFs like GE, which also add organic matter to the soil.

While BC soil amendments may generally enhance soil health and plant growth in forestry, agriculture, and other applications, responses will vary because BCs differ and are influenced by soil type, climate, vegetation, and management [54]. Agriculture is using BC to improve soil bulk density, root penetration, aggregate stability, water infiltration, water holding capacity or retention, nutrient leaching, pore distribution, organic matter, carbon sequestration, toxins and pollutants, soil disease pathogens, beneficial nematodes, nitrogen-mineralization rate and microbial biomass, respiration rate, and genetic diversity [55].

BC may remediate contaminated soils [56], restore degraded land, and increase agriculture efficiency and carbon fixation [57]. In Brazil, adding 4.2 t/ha/year of sugarcane BC in sugarcane fields could increase soil C by 2.35 t C/ha/year [58]. In European agriculture, BC + low input of nitrogen fertilizer provided the highest C sequestration (61.1 t CO2e/t of biomass) [59]. The renewed interest in biochar was stimulated by the discovery of high organic carbon and remarkably fertile soils in South America, especially Amazonia, that have been called "Amazonian Dark Earths or Terra Preta de Indio" (black Earth of Indians). These soils maintain fertility for years. Remarkably, these areas of the world are often characterized by low fertility and nutrient holding capacity. The fertility of the Amazonian Dark Earths is believed to be largely a consequence of charcoal/biochar applications by the indigenous tribes of the region and the benefits in the soils persisted for thousands of years.

BC is produced via pyrolysis, that is, heating wood in a very low oxygen environment to remove all moisture and volatiles, maximize carbon content, and minimize ash content while increasing porosity and maximizing surface area. BC pyrolysis technologies range from simple batches production techniques, such as open pits, mounds, and kilns, to continuous production systems using rotary kilns and retorts [7].

Given the trends toward sustainable business models and reducing the CO_2 footprint of production systems, the type of technology employed is an important consideration in BC production. As one moves up the technology scale, BC producers have the ability to control greater portions of the production process. A simple batch technology has limited ability to control the pyrolysis process compared to continuous production systems. Some of the operating metrics producers may want to control pyrolysis temperature, residence time, combustion of volatiles, and energy capture. To sustainably produce BC, operators will want to control all of these items and more, including, emissions and the source of feedstock.

While there is value in producing BC in remote areas to help support local agriculture or possibly even for export, many of these operations are not sustainable supply chains over the long term. The least sustainable producers are where the virgin forest is harvested to produce BC in open pits, mounds, or kilns. To truly be sustainable, pyrolysis operations should capture all components of value including fully combusting the volatiles inherent in the feedstock, converting this to a usable form of what is bioenergy, and then utilizing that energy in other applications (**Figure 4**). GCS is committed to these goals and the sustainable production of BC.

GCS' operations capture and utilize all components of value in BC production. With a commitment to sustainability and to further improve efficiency, GCS has designed its pyrolysis operations to be continuous, minimize the use of electricity, and capture and convert all volatiles into usable forms of energy for other applications. With a sustainable BC production process, carbon sequestered will have a greater beneficial impact.

Interest in and demand for BC documented in 2020 [7] are still growing due to improved BC production techniques, but BC's multiple applications vary widely in potential market size, timing, competitiveness, and pricing compared to alternative products (**Table 15**). With the need to replace the substantial loss of soil carbon due to modern agricultural practices [60] and considering the emerging carbon cascades [61], the applications and future potential markets become quite large. There are growing opportunities to utilize BC for (1) soil nutrient and water retention, (2) remediation of contaminated soils and water, (3) filler in concrete, asphalt, and tires, (4) acoustic and thermal insulation in walls, ceilings, and floors, (5) carbon fibers

Figure 4.
GCS' pyrolysis process with integrated heat capture and utilization.

Application	Market	Timing	Competition	Pricing
Soil carbon	Large	Current	Growing	Low
Specialty soil	Moderate	Emerging	Moderate	Moderate/high
Crop yield	Moderate	Current	High	Low/moderate
Carbon sequestration	Very large	Emerging	Moderate	Moderate
Nutrient retention	Large	Current	Moderate	Moderate
Water retention	Large	Current	Moderate	Moderate
Water purification	Large	Emerging	Low	High
General industrial	Large	Current	Moderate	Moderate
Specialty industrial	Moderate	Emerging	Low	High

Table 15.
Relative market, timing, competition, and pricing for BC applications.

and polymers, (6) protection against electrosmog, (7) filtration media, and (8) heavy metal adsorption. Growing trends in developing sustainable supply chains and reducing societal carbon footprint will help accelerate the growth of many of these markets.

5. Conclusions

Estimated carbon sequestration by *Eucalyptus* in Florida can be sizeable but depends on site conditions and management options. *Eucalyptus* managed in long rotations for mulch wood production sequesters less but still significant amounts of carbon. *Eucalyptus* cultivars are responsive to intensive culture in SRWC systems that may economically produce high-quality BC, which in turn can be a useful soil amendment for their culture and increase total carbon sequestration. In evaluating the tradeoffs of alternative management options to intensive SRWC culture, growers should consider soil type, planting density, and soil amendments. Amending soil with BC can both increase and accelerate total carbon sequestration and also help offset any carbon loss that takes place in growing *Eucalyptus*. Demand for sustainably produced BC is growing due to its multiple applications beyond soil carbon sequestration.

Acknowledgements

The authors gratefully acknowledge the direct and/or indirect support provided by the IRREC, GCS, Green Technologies, Evans Properties, US EcoGen, Becker Tree Farm, Water Conserv II, ArborGen, the EREC, the TPA, J and B Rodriguez, J Phillips, and J Rockwood. This research did not receive any specific grant from funding agencies in the public, commercial, or not-for-profit sectors.

Conflict of interest

The authors declare no conflict of interest.

Author details

Donald L. Rockwood[1]*, Kyle W. Fabbro[2], Martin F. Ellis[3], Monica Ozores-Hampton[4] and Amir Varshovi[5]

1 Florida FGT LLC, UF/IFAS School of Forest, Fisheries, and Geomatics Sciences, Gainesville, FL, USA

2 Lykes Brothers, Palmdale, FL, USA

3 Green Carbon Solutions (GCS), Pepper Pike, OH, USA

4 Terranutri, Ave Maria, FL, USA

5 Green Technologies, Gainesville, FL, USA

*Address all correspondence to: dlrock@ufl.edu

References

[1] CIRAD—FRA, IUFRO—AUT, MUSE—FRA. Eucalyptus 2018: Managing Eucalyptus Plantation under Global Changes. Montpellier, France: 2018

[2] Diaz-Balteiro L, Rodriguez LCE. Optimal rotations on Eucalyptus plantations including carbon sequestration—A comparison of results in Brazil and Spain. Forest Ecology and Management. 2006;**229**(1-3):247-258. DOI: 10.1016/j.foreco.2006.04.005

[3] Zhou X, Wen Y, Goodale UM, Zuo H, Zhu H, Li X, et al. Optimal rotation length for carbon sequestration in Eucalyptus plantations in subtropical China. New Forests. 2017;**48**:609-627. DOI: 10.1007/s11056-017-9588-2

[4] Nawaz MF, Shah SAA, Gul S, Afzal S, Ahmad I, Ghaf A. Carbon sequestration and production of Eucalyptus camaldulensis plantations on marginal sandy agricultural lands. Pakistan Journal of Agricultural Sciences. 2017;**54**(2):335-342. DOI: 10.21162/PAKJAS/17.4432

[5] Kumari P, Mishra AK, Kumar M, Chaudhari SK, Singh R, Singh K, et al. Biomass production and carbon sequestration of Eucalyptus tereticornis plantation in reclaimed sodic soils of northwest India. Indian Journal of Agricultural Sciences. 2019;**89**(7):1091-1095

[6] Arroja L, Dias AC, Capela I. The role of Eucalyptus globulus forest and products in carbon sequestration. Climatic Change. 2006;**74**:123-140. DOI: 10.1007/s10584-006-3461-1

[7] Rockwood DL, Ellis MF, Liu R, Zhao F, Fabbro KW, He Z, et al. Forest trees for biochar and carbon sequestration: Production and benefits. In: Abdel-hafez A, Abbas M, editors. Applications of Biochar for Environmental Safety. London, UK: IntechOpen; 2020

[8] Fabbro KW, Rockwood DL. Optimal management and productivity of *Eucalyptus grandis* on former phosphate mined and citrus lands in central and southern Florida: Influence of genetics and spacing. In: Proceedings 18th. Biennial Southern Silvicultural Research Conference, March 2-5, 2015, Knoxville, TN. e-Gen. Tech. Rpt. SRS-212; 2016. pp. 510-517. Available from: http://www.srs.fs.usda.gov/pubs/gtr/gtr_srs212.pdf

[9] Ahmed A, Kurian J, Raghavan V. Biochar influences on agricultural soils, crop production, and the environment: A review. Environmental Reviews. 2016;**24**(4):495-502

[10] Jeffery S, Abalos D, Prodana M, Bastos AC, van Groenigen JW, Hungate BA, Verkeijen F. Biochar boosts tropical but not temperate crop yields. Environmental Research Letters. 2017;12 05 3001

[11] Ahmed F, Arthur E, Plauborg F, Razzaghi F, Korup K, Andersen MN. Biochar amendment of fluvio-glacial temperate sandy subsoil: Effects on maize water uptake, growth and physiology. Journal of Agronomy and Crop Science. 2018;**204**(2):123-136

[12] Blanco-Canqui H. Biochar and soil physical properties. Soil Science Society of America Journal. 2017;**81**(4):687-711

[13] Bruun EW, Petersen CT, Hansen E, Holm JK, Hauggaard-Nielsen H. Biochar amendment to coarse sandy subsoil improves root growth and increases water retention. Soil Use and Management. 2014;**30**(1):109-118

[14] Hussein H, Farooq M, Nawaz A, Al-Sadi AM, Solaiman ZM, Alghamdi SS, et al. Biochar for crop production: Potential benefits and risks. Journal of Soils and Sediments. 2017;**17**(3):685-716

[15] Rockwood DL, Ellis MF, Fabbro KW. Economic potential for carbon sequestration by short rotation eucalypts using biochar in Florida, USA. Trees, Forests and People. 2022. DOI: 10.1016/j.tfp.2021.100187

[16] Campoe OC, Stape JL, Laclau J-P, Marsden C, Nouvellon Y. Stand-level patterns of carbon fluxes and partitioning in a Eucalyptus grandis plantation across a gradient of productivity in Sao Paulo, Brazil. Tree Physiology. 2012;**32**:696-706. DOI: 10.1093/treephys/tps038

[17] Langholtz MH. Economic and environmental analysis of tree crops on marginal lands in Florida [PhD dissertation]. University of Florida; 2005. Available from: https://ufdc.ufl.edu/UFE0012141/00001

[18] Goncalves JLM, Stape JL, Laclau J-P, Bouillet J-P, Ranger J. Assessing the effects of early silvicultural management on long-term site productivity of fast-growing eucalypt plantations: The Brazilian experience. Southern Forests. 2008;**70**(2):105-118

[19] du Plessis M, Kotze H. Growth and yield models for Eucalyptus grandis grown in Swaziland. Southern Forests: A Journal of Forest Science. 2011;**73**(2):81-89. DOI: 10.2989/20702620.2011.610873

[20] Shackley S, Sohi S, Ibarrola R, Hammond J, Masek O, Brownsort P, et al. Biochar tool for climate change mitigation and soil management. In: Meyers RA, editor. Encyclopedia of Sustainability Science and Technology. New York, NY: Springer; 2012. DOI: 10.1007/978-1-4419-0851-3

[21] Meskimen G, Franklin EC. Spacing Eucalyptus grandis in southern Florida: A question of merchantable versus total volume. Southern Journal of Applied Forestry. 1978;**1**:3-5

[22] Pisano SM, Rockwood DL. Stormwater phytoremediation potential of Eucalyptus. In: Proceedings 5th Biennial Stormwater Research Conference; Nov. 5-7, 1997; Tampa, FL. Brooksville, FL: Southwest Florida Water Management District; 1997. pp. 32-42

[23] Rombel A, Krasucka P, Oleszczuk P. Science of the total environment. Sustainable Biochar-Based Soil Fertilizers and Amendments as a New Trend in Biochar Research. Elsevier. Available from: https://www.sciencedirect.com/science/article/pii/S0048969721066663

[24] Chen L, Li W, Xiao Y. Ecological Indicators, "Biochar and Nitrogen Fertilizer Increase Glomus Synergism and Abundance and Promote *Trifolium Pratense* Growth While Inhibiting Pollutant Accumulation." Elsevier. Available from: https://www.sciencedirect.com/science/article/pii/S1470160X21010426

[25] Das SK, Ghosh GK. Developing biochar-based slow-release N-P-K fertilizer for controlled nutrient release and its impact on soil health and yield. Biomass Conversion and Biorefinery. 2021. Available from: https://link.springer.com/10.1007/s13399-021-02069-6

[26] Dong L et al. Biochar and Nitrogen Fertilizer Co-Application Changed SOC Content and Fraction Composition in Huang-Huai-Hai Plain. China: Elsevier; 2021

[27] Mendes JDS, et al. Effect of poultry litter biochar on the nutritional status

of corn. SciELO Brasil. Available from: https://www.scielo.br/j/rcaat/a/Z8bgwrqVZ9rYd7TysWq9P7k/abstract/?lang=en

[28] Alferez F. Compost utilization in fruit crops. In: Ozores-Hampton M, editor. Compost Utilization in Production of Horticultural Crops. Boca Raton, USA: Taylor & Francis Group, LLC; 2021. pp. 51-58

[29] Ozores-Hampton M, Stansly P. Using compost in citrus. Citrus Magazine. 2015;**2015**:8-11. Available from: https://crec.ifas.ufl.edu/extension/trade_journals/2015/2015_December_compost.pdf

[30] Ozores-Hampton M. Using organic amendments in citrus production. In: 14th U.S. Composting Council Ann. Conf. & Tradeshow. Book of Abstr. 2006. p. 43

[31] Litvany M, Ozores-Hampton M. Compost use in commercial citrus in Florida. HortTechnology. 2002;**12**:332-335

[32] Obreza TA, Ozores-Hampton M. Management of organic amendments in Florida citrus production systems. Soil Crop. 2000;**59**:22-27

[33] Ozores-Hampton M. Compost utilization in vegetable crops. In: Ozores-Hampton M, editor. Compost Utilization in Production of Horticultural Crops. Boca Raton, USA: Taylor & Francis Group, LLC; 2021. pp. 59-76

[34] Ozores-Hampton M. Impact of compost in soil health. In: Ozores-Hampton M, editor. Compost Utilization in Production of Horticultural Crops. Boca Raton, USA: Taylor and Francis Group, LLC; 2021. pp. 9-26

[35] Christie S, Scholes R. Carbon storage in Eucalyptus and pine plantations in South Africa. Environmental Monitoring and Assessment. 1995;**38**:231-241

[36] Du H, Zeng F, Peng W, Wang K, Zhang H, Liu L, et al. Carbon storage in a Eucalyptus plantation chronosequence in southern China. Forests. 2015;**6**:1763-1778. DOI: 10.3390/f6061763

[37] Zhang H, Guan D, Song M. Biomass and carbon storage of Eucalyptus and Acacia plantations in the Pearl River Delta, South China. Forest Ecology and Management. 2012;**277**:90-97. DOI: 10.1016/j.foreco.2012.04.016

[38] Ulman Y, Avudainayagam S. Carbon storage potential of Eucalyptus tereticornis plantations. Indian Forester. 2014;**140**(1):53-58

[39] Rockwood DL, Peter GF. Eucalyptus and Corymbia species for mulchwood, pulpwood, energywood, bioproducts, windbreaks, and/or phytoremediation. Florida Cooperative Extension Service Circular. 2018;**1194**:6

[40] Rockwood DL. History and status of Eucalyptus improvement in Florida. In: Naik A, Ayeni LS, editors. New Perspectives in Agriculture and Crop Science. Vol. Volume 3. Book Publisher International; 2020

[41] Stavins RN, Richards KR. The cost of U.S. forest-based carbon sequestration. Pew Center on Global Climate Change. 2005. Available from: https://www.c2es.org/document/the-cost-of-u-s-forest-based-carbon-sequestration/

[42] California Air Resources Board. Compliance Offset Protocol U. S. Forest Offset projects. Available from: arb.ca.gov/cc/capandtrade/protocols/usforest/usforestprojects_2015.htm

[43] Lehman J, Joseph S. Biochar for Environmental Management. New York: Earth Scan; 2006. pp. 188-200

[44] Bis Z, Kobyłecki R, Ścisłowska M, Zarzycki R. Biochar—Potential tool to combat climate change and drought. Ecohydrology & Hydrobiology. 2018;**18**(4):441-453

[45] Rouse R, Ozores-Hampton M, Roka F, Roberts P. Rehabilitation of huanglongbing infected citrus trees using severe pruning and foliar nutritionals. HortScience. 2017;**52**:972-978

[46] Goudiaby A, Diedhiou S, Diatta Y, Adiane A, Diouf P, Fall S, et al. Soil properties and groundnut (*Arachis hypogea* L.) responses to intercropping with Eucalyptus camaldulensis Dehn and amendment with its biochar. Journal of Material Environment Science. 2020;**11**(2):220-230

[47] Coumaravel, K. Effect of cotton stalk biochar on maize productivity under calcareous clay soil condition. The-pharmajournal.com. Available from: https://www.thepharmajournal.com/archives/2020/vol9issue10/PartG/9-9-88-850.pdf

[48] Agbede TM, Adekiya AO. Influence of biochar on soil physicochemical properties, erosion potential, and maize (*Zea mays* L.) grain yield under sandy soil condition. Communications in Soil Science and Plant Analysis. 2020

[49] Sajedi A, Sajedi NA. Effect of application biochar and priming and foliar application with water and salicylic acid on physiological traits of dry land safflower. 2020. Available from: https://agris.fao.org/agrissearch/search.do?recordID=IR2020700038

[50] Ali I, Ullah S, He L, Zhao Q, Iqbal A, Wei S, et al. Combined application of biochar and nitrogen fertilizer improves rice yield, microbial activity and N-metabolism in a pot experiment. Peer Journal Communication. 2020;**8**:e10311. DOI: 10.7717/peerj.10311

[51] Waqqas KTM, Fan L, Cai Y, Tayyab M, Chen L, He T, et al. Biochar amendment regulated growth, physiological, and biochemical responses of conifer in red soil. iForest-Biogeosciences. 2020. Available from: Sisef.It. http://www.sisef.it/iforest/contents/?id=ifor3416-013

[52] Pan L, Xu F, Mo H, Corlett RT, Sha L. The potential for biochar application in rubber plantations in Xishuangbanna, Southwest China: A pot trial. Biochar. 2020. DOI: 10.1007/s42773-020-00072-0

[53] Sánchez-Monedero M, Cayuela M, Sánchez-García M, Vandecasteele B, D'Hose T, López G, et al. Agronomic evaluation of biochar, compost and biochar-blended compost across different cropping systems: Perspective from the European Project FERTIPLUS. Agronomy. 2019;**9**:225. DOI: 10.3390/agronomy9050225

[54] Brockamp, R.L, Weyers, S.L. Chapter 8—Biochar amendments show potential for restoration of degraded, contaminated, and infertile soils in agricultural and forested landscapes. In: Soils and Landscape Restoration. Elsevier; 2021. Available from: https://www.sciencedirect.com/science/article/pii/B9780128131930000084.

[55] Yang D, Yunguo L, Shaobo L, Zhongwu L, Xiaofei T, Xixian H, et al. Biochar to improve soil fertility. A review. Agronomic Sustainable Development. 2016;**36**:36

[56] Papageorgiou A, Azzi ES, Enell A, Sundberg C. Biochar produced from wood waste for soil remediation in Sweden: Carbon sequestration and other environmental impacts. Science of The Total Enviroment. 2021;**776**:145953

[57] Dar AA, Mohd YR, Javid M, Waseem Y, Khursheed AW, Dheeraj V.

Biochar: Preparation, properties and applications in sustainable agriculture. International Journal of Theoretical & Applied Sciences. 2019;**11**(2):29-40

[58] Lefebvre D, Williams A, Meersmans J, Kirk GJD. Modelling the potential for soil carbon sequestration using biochar from sugarcane residues in Brazil. Scientific Reports. 2020. Available from: Nature. Com. https://www.nature.com/articles/s41598-020-76470-y

[59] Solinas S, Tiloca MT, Deligios PA, Cossu M, Ledda L. Carbon footprints and social carbon cost assessments in a perennial energy crop system: A comparison of fertilizer management practices in a Mediterranean area. In: Agricultural Systems. Elsevier; 2020

[60] Montgomery D. Growing A Revolution: Bringing Our Soil Back to Life. New York: W Norton and Company; 2017

[61] Bales A, Draper K. Using Fire to Cool the Earth. London: Chelsea Green Publishing; 2018

Chapter 3

Regenerating Soil Microbiome: Balancing Microbial CO_2 Sequestration and Emission

Mohd N.H. Sarjuni, Siti A.M. Dolit, Aidee K. Khamis, Nazrin Abd-Aziz, Nur R. Azman and Umi A. Asli

Abstract

Soil microbiome plays a significant role in soil's ecosystem for soils to be physically and biologically healthy. Soil health is fundamental for plant growth and crops productivity. In the introduction part, the roles and dynamics of the microbial community in soils, primarily in the cycle of soil organic carbon and CO_2 release and absorption, are deliberated. Next, the impact of crop management practices and climate change on the soil carbon balance are described, as well as other issues related to soil degradation, such as unbalanced nutrient recycling and mineral weathering. In response to these issues, various approaches to soil regeneration have been developed in order to foster an efficient and active soil microbiome, thereby balancing the CO_2 cycle and carbon sequestration in the soil ecosystem.

Keywords: soil microbiome, soil health, microbial CO_2, CO_2 sequestration, CO_2 emission

1. Introduction

Microbes are the most diverse organisms on the planet, both in terms of species and in terms of driving vital Earth system operations like the carbon cycle. The majority of this microbial biodiversity is found in soils [1]. According to Lederberg and McCray [2], the term microbiome refers to "the biological community of commensal, symbiotic, and pathogenic microbes that share human body space." This term grew in popularity as its definition evolved from organisms as taxonomic units (i.e., microbiota) to a collective genetic material throughout the years. However, as the term's popularity grew, there are various definitions of the term microbiome in the scientific literature.

Nowadays, most "microbiome" research focuses solely on bacteria, and the term "microbiome" is used interchangeably with "bacteria." As a result, new words for various microbial groupings have emerged, such as mycobiome, which refers to fungi, virome for the viruses, and eukaryome for the microbial eukaryotes [3]. Furthermore, the composition of microbiomes is known to change across time and space, making it difficult to find consistent and dependable sources of specific microbiomes [4].

The microbiome of the Earth accounts for almost half of all biomass on the globe [1]. Recent advances in DNA sequencing techniques have expanded our understanding of microbial biogeography, particularly among bacteria and fungi [5, 6]. Currently, the diverse composition of soil microbial community is widely known worldwide. The soil microbiome governs the biogeochemical cycling of macronutrients, micronutrients, and other elements that are vital for plants growth and animal life.

Microbiomes play an important role in a variety of biogeochemical processes, including the carbon and nitrogen cycles, which are necessary for ecosystems to function properly and sustainably. What functions do bacteria play in nutrient cycling and carbon sequestration to support the forests? It is critical to investigate the dynamics of microbial communities in order to comprehend their vital function in such a unique ecosystem. Acknowledging microbial ecology will aid in their management practices and protection, allowing peat accretion to continue and their carbon sequestration capacity to be protected [7].

The significance of soil microbiome activity in the soil ecosystem dynamics demands special consideration, as it promotes soil health and plant productivity [8]. Soil microbial activity is a possible indicator of soil quality as it responds quickly to changes in soil management and the environment. The carbon in crop residues moves *via* soil microbial biomass at least once, where it is moved from one C pool to another and eventually lost as carbon dioxide (CO_2) [9]. It is critical to understand the factors that determine the richness of soil bacterial communities, as well as the organization of these communities, in order to forecast the responses of an ecosystem toward a specific environment. Changes in microbial populations or activity can occur prior to visible changes in soil physical and chemical properties, acting as an early indicator of soil improvement or degradation [10].

Soil characteristics such as pH, carbon, and nitrogen have been shown to influence soil microbial diversity and biogeography [11]. As a result, changes in the structure and behavior of soil microbial communities are more likely to be caused by differences in soil characteristics. Aside from that, soil organic matter (SOM) is critical to the function and quality of the soil. The high amount of SOM could increase nutrient availability while also improving the physical and biological features of the soil [12]. The level of soil organic carbon (SOC) is used to quantify the amount of SOM, and changes in SOC have an impact on the carbon (C) and nitrogen (N) cycles in terrestrial ecosystems [13]. The combined effects of chemical and biological features of the soil will affect the organic C and N fractions in organic compounds. As a result, understanding the processes that determine soil fertility, which is critical in farmland production systems, requires knowledge of soil microbial community dynamics and the factors that influence those dynamics in croplands.

2. CO_2 balance in soils

The technique of increasing soil carbon storage by reducing net CO_2 emissions in agricultural soils is known as carbon sequestration. Soil carbon sequestration (SCS) is the process of absorbing C-containing compounds from the atmosphere and storing them in soil C pools. Variations in the ability to store carbon in soils have been linked to the activity of the soil microbial community (SMC). The turnover and supply of nutrients, as well as the rate of decomposition of SOM, are all influenced by the structure and activity of the SMC, which is crucial for the maintenance of soil ecosystem services. As a result, the influence of farming activities on SMC and SCS should be quantified as part of any soil management practice's sustainability evaluation.

Because a big fraction of the biomass is produced in agricultural systems cycles *via* the soil decomposer community, the quantity of gross CO_2 fluxes between agricultural soils and the atmosphere is significant. The difference between photosynthetically fixed CO_2 entering the soil as plant wastes and CO_2 exhaled during decomposition, on the other hand, is far smaller. This distinction determines whether the ecosystem is a CO2 source or sink in terms of its net carbon balance.

Raising the C content of agricultural soils is a well-known technique. The equilibrium between C inputs from plant residues and C losses, primarily through decomposition, determines the soil C levels. The increasing residue inputs and/or delaying breakdown rates (i.e., heterotrophic soil respiration) also govern the C level in soils. The relationship between C inputs and SOC levels could be straightforward; in which many agricultural soils' steady-state C contents have been shown to be linearly related to C input levels, that is compatible with the current SOM dynamics theory [14]. This may not be the case in soils with exceptionally high quantities of carbon, which may exhibit "saturation" behavior.

The following factors must be considered when developing soil carbon sequestration management practices and policies: Soils have a finite capacity to store carbon, gains in soil carbon can be reversed if proper management is not maintained, and fossil fuel inputs for various management practices must be factored into the total agricultural CO_2 balance [15].

The interaction of numerous ecosystem activities, the most important of which are photosynthesis, respiration, and decomposition, results in the SOC level. Photosynthesis is the process of converting atmospheric CO_2 into plant biomass. The root biomass of a plant determines the majority of SOC ingestion rates, however, litter deposited by plant shoots also plays a role. The growth and death of plant roots, as well as the transfer of carbon-rich molecules from roots to soil microbes, produce carbon in the soil both directly and indirectly.

Decomposition of biomass by soil microbes leads to carbon loss as CO_2 as a result of microbial respiration. Through the formation of humus, a material that gives carbon-rich soils their unique black hue, a small fraction of the original carbon is kept in the soil (**Figure 1**). These various forms of SOC differ in their recalcitrance, or resistance to decomposition. Humus is a recalcitrant plant that takes a long time to degrade, resulting in a long period of time spent in the soil. Plant waste is less abrasive; therefore, it stays in the soil for a shorter period of time. When carbon imports and outputs are in equilibrium, there is no net change in SOC levels. When carbon inputs from photosynthesis exceed carbon losses, SOC levels rise over time.

2.1 Impact of climate change on soils carbon

The effects of climate change on soil functions, including soil carbon, is a complex subject since numerous direct and indirect factors are involved. For instance, the atmospheric temperature may affect the rate of SOM decomposition, a process that could release greenhouse gases that contribute to climate change [16]. The effects of moisture and temperature due to climate change will be highlighted as key parameters since soil humidity and temperature are among the most important variables in determining microbial activity and therefore SOC [17].

One of the most critical effects of climate change on soil is the alteration of rainfall patterns, resulting in intense rain and drought. These phenomena may be beneficial

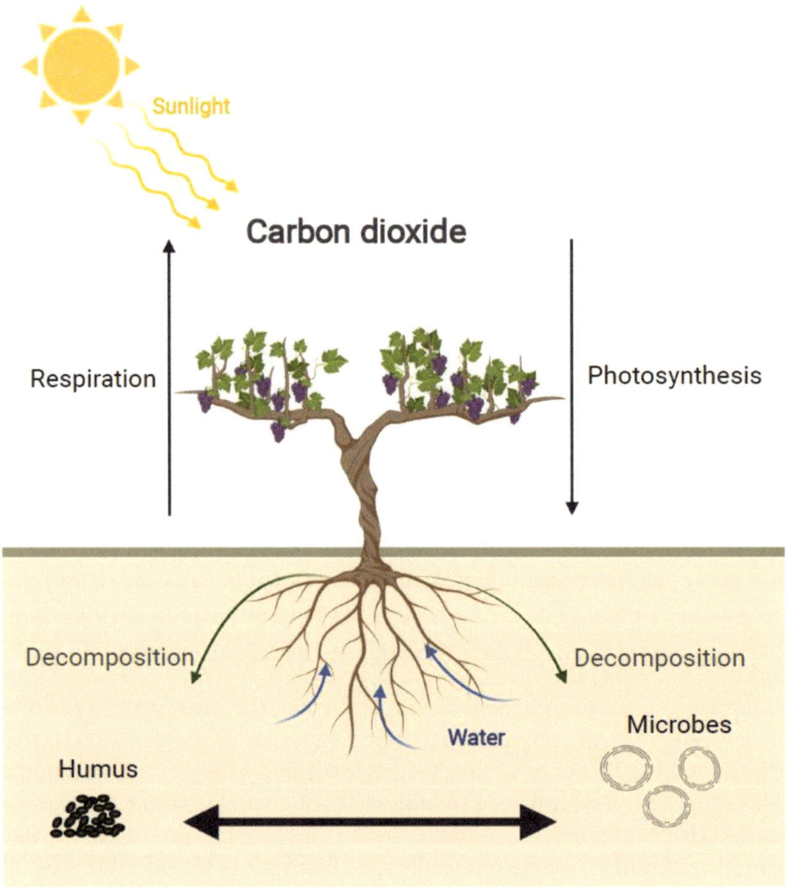

Figure 1.
Carbon inputs from photosynthesis and carbon losses from respiration govern the carbon balance within the soil. Humus, long-lived storage of SOC, is formed through the decomposition of roots and root products by soil bacteria. Created with BioRender.com.

or detrimental according to the agricultural activities and climatic requirements, but they present economic challenges nonetheless [18]. The selective migration of soil particles, where fine particles and micro-aggregates are transported *via* erosion while macroaggregates are left *in situ* resulted in different carbon mineralization patterns. These lateral redistributions of sediments create (i) eroded environments dominated by large particles exhibiting increased porosity and permeability but decreased water-holding capacity, and (ii) deposited environments where enrichment of fine particles enhances the water holding capacity [17]. Similar to water erosion induced by water runoff, wind erosion induced by drought also redistributes a large amount of SOC as well as soil inorganic carbon (SIC). In addition to soil particles, the net effects of these soil C redistribution on the soil as a C source or sink also depend on site-specific topography (such as slope gradient and location), distribution distance, and duration [19]. Some of these interacting factors are outlined in **Figure 2**.

Among the most consistent narrative of climate change is climate warming as a result of rising temperature [20]. Climate warming has been associated mainly with SOC decomposition due to the effects of temperature on soil microbial community and their enzymatic and metabolic activities. Unlike the effects of moisture, however, the

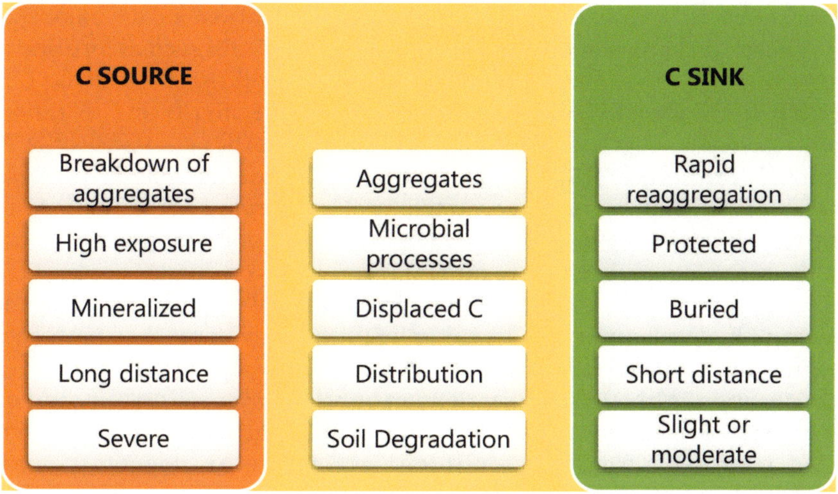

Figure 2.
Interaction of diverse factors affecting soil as C source or sink.

dynamic relationship between temperature and soil C is less certain and more constrained. In general, elevated atmospheric temperature could also elevate soil temperature, which would subsequently elevate microbial processes and SOC decomposition rate [21]. This is not always the case due to the difference in temperature sensitivity of soil biota, especially the microbial community, where the higher-temperature sensitivity such as in colder regions exhibited more enhanced soil respiration, potentially resulting in a net efflux of C toward increased atmospheric CO_2 in comparison with those inhabiting soils in hotter regions [22]. In contrast, a higher rate of microbial OM decomposition was reported in hotter regions, suggesting other environmental factors that may affect the SOC, including topography, soil texture, and pH. Ultimately, climate warming leads to decreased SOC input and increased SOC output [23].

2.2 Effect of agronomic management on soils microbiome and CO_2 balance in soils

Agronomic management involves a combination of soil and crop management practices that when appropriately applied will improve soil performance and nutrient availability, and contribute to better growth and higher crop yield [24]. These management practices can be further categorized into an untargeted approach based on common agricultural practices, or targeted approaches based on specific interactions between soil and plant. Targeted approaches often involve biotechnological applications such as biofertilizers and biostimulants. Regardless of the type of approach, the soil microbiome will be affected either directly or indirectly. Considering that the soil microbiome is regarded as the primary organism that may influence the overall plant health due to its close interaction with plant roots, applying the right management practice is crucial toward achieving the goal of food security for the growing global population [25]. Therefore, soil microbiome must not be overlooked in agronomic management practice especially when SOM is concerned due to its major role in soil C pool. For instance, additional OM applications may result in increased decomposition and reduced C storage due to reduced microbial C use efficiency, positive priming effect from enhanced mineralization of SOM, as well as increased C skimming due to accumulation of microbial products and residues, or necromass over time [26].

SOC is known to be directly influenced by the stabilization and decomposition of SOM. Therefore, agronomic management that boosts SOM such as fertilization, conservation tillage, cover cropping, and crop rotation will also affect SOC [27]. More importantly, soil biotic and abiotic factors such as texture, moisture, C/N ratio, SOC content, pH, climate, vegetation, and land use also affect the persistence of SOM, and ultimately the C pools [28]. It is due to the complex interactions of these various factors that it is uncommon for an ecosystem to change from a net C source to a C sink in a relatively short time [29]. Thus, agronomic management practice must take into account the most appropriate way to minimize its impact on climate change [27].

Crop management refers to a collection of agricultural activities aimed at enhancing crop growth, development, and production. It starts with seedbed preparation, seed sowing, and crop maintenance and concludes with crop harvest, storage, and commercialization. Although fertilization not only improves soil fertility and quality but also crop production, it causes soil pollution, soil hardness, organic matter mineralization, increased nitrous oxide emissions, and nitrate leaching into groundwater and surface waters [30]. Fertilizer application considerably affected the soil C/N ratio. When Liu et al. [30] analyzed chemical and organic fertilizers, they discovered that chemical fertilizer (NPK) treatment lowers soil pH, and when combined with organic fertilizer, it lowers the soil pH even more. Furthermore, the relative populations of microbiome components varied after organic waste (straw) treatment due to changes in ammonium nitrogen (NH_4^+-N) and nitrate nitrogen (NO_3^+-N).

Bhattacharyya et al. [26] reported that the influence of organic matter accessibility on the significance of SMC to soil C control can be explained in numerous ways:

1. Increased organic matter inputs may hasten decomposition and decrease C storage by reducing microbial C usage efficiency.

2. Greater organic matter additions, labile carbon inputs, or nutrient inputs result in increased SOM mineralization in soil, which is referred to as a positive priming effect.

3. Increased organic matter additions can boost C skimming by increasing the formation of microbial necromass over time.

The interrelationship between nutrients, roots, water, and SOM is another component that influences SOM to build up in more complex cropping systems. In the surface soil layer, available nutrients are dynamic; they may be reduced by net microbial immobilization during heavy litter intake times and abundant during times of net mineralization. Microorganisms regulate root proliferation through their effects on nutrient availability and water, while roots influence microbial activity through their effects on nutrient availability. Increased litter inputs encourage competition for nutrients between microorganisms. When litter and organic matter pool sizes increase over longer periods, mechanisms favoring C sequestration are reinforced such as improved plant water availability, stronger nutrient recycling capacity, and reduction of nutrient leakage. Since microbial and plant respiratory processes are dominated by nutrient availability, cover crops that increase CO_2 and N_2O fluxes would have a good impact on soil respiration.

The pH of the soil influences microbial activity. As a result, soil management activities such as liming have an impact on soil emissions as additional carbonate can be emitted as CO_2. Soil emissions are reduced when the soil is acidic. The ideal pH for methanogenesis (CH_4 generation) is found between pH 4 and 7. CO2 emissions are at their highest when the pH levels are neutral. Under acidic soil conditions, N2O emissions are reduced. Because the balance between NH3 and NO3 flips to ammonia at higher pH values, nitrification rises. However, there was no evidence of a link between NO and N2O emissions and pH. Denitrification produces NO emissions under acidic soil conditions, whereas nitrification produces NO emissions under alkaline soil conditions.

Crop rotation (CR) changes soil microbial profiles toward microorganisms with C-sequestering characteristics. According to Venter et al. [31], microbial diversity and richness can be increased by 15 and 3.4%, respectively, using CR. Different crop rotation practices may cause variations in soil C storage and SMC use. After a long-term CR practice involving legumes, SOC stock, MBC, and soil enzymatic activity (acid/alkaline phosphatase, beta-glucosidase, and arylsulfatase) may rise. The presence of legumes in CR may help to protect the SMC in general.

3. Balancing soil CO_2: a race against time

If left undisturbed, soil carbon may remain sequestered for thousands of years [32]. Disturbed soils, which are primarily due to intensive cultivation, have decreased the soil's ability to maintain and store carbon, amplifying the impacts of climate change and the accompanying costs to mitigate them [33]. While soil ability as a carbon sequester varies with location, climate, and soil type, one common cause of carbon loss, the majority of which is emitted as carbon dioxide is due to unsustainable management practices at the macroscopical level. Further approaches to sustainable management practices should consider and employ our current knowledge at the microscopical or cellular level. Acknowledgment and immediate actions from all relevant stakeholders must be engaged in the race against time to mitigate the climate change while ensuring the benefits for the environment, community, and economy.

3.1 Macroscopical level: sustainable soil management

Among the easiest options to avoid or reduce soil carbon loss are sustainable soil management practices at the ground level where the results of carbon sequestration can be detected within several years of implementation [34]. Enhanced food security and nutrition as well as improved ecosystem services are some of the possible benefits to be gained over the short to medium term (**Figure 3**).

Sustainable soil management practices involve the increase of SOM to offset the effects of land conversion, tillage disturbance, soil erosion, and leaching from human activities [35]. The conundrum in sustainable soil management practices is that determining the best practices does not only depend on the dynamic properties of the soil, but also relies on various environmental conditions and social and economic factors. Nevertheless, several studies agreed that sustainable soil management practices should include the following:

i. Adoption of no-till or conservation tillage to preserve soil structure [36];

ii. Use of cover crops to increase SOM, water holding capacity, and protection from wind and water erosion [37];

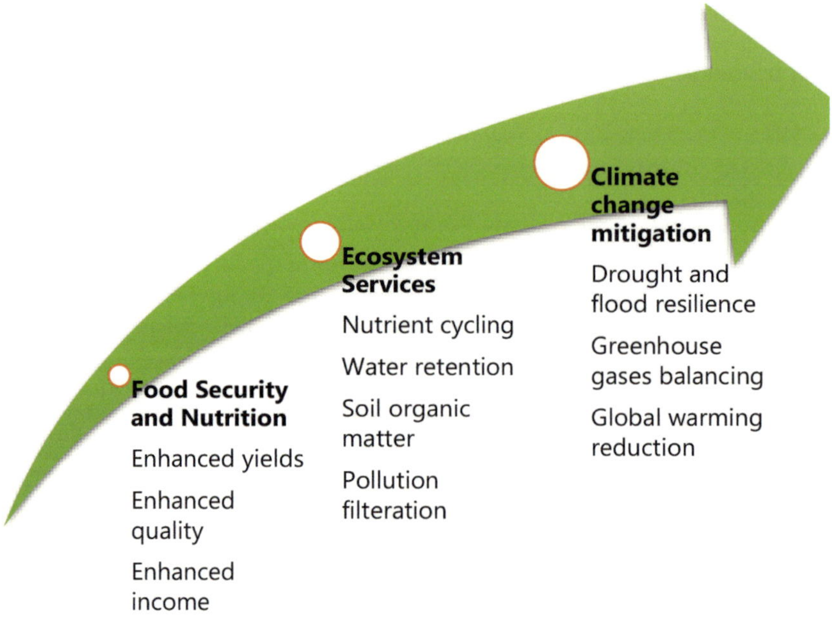

Figure 3.
Benefits of sustainable soil management practices.

iii. Organic soil amendment from plant residues, compost, and biochars to lower C:N ratio [38]; and

iv. Better irrigation to manage soil salinity [39].

Following these strategies, the measurement of soil CO_2 flux can be used to determine whether the ecosystem is functioning as a net carbon sink [40]. This is important since there are reports that the application of organic manures and residues could increase CO_2 emission, which negates the goals of mitigating climate change [41]. Verifying the most appropriate sustainable soil management practices is deemed of the utmost importance to ensure successful soil-specific microbial carbon sequestration.

3.2 Microscopical level: engineered microorganisms

Modern climate change mitigation techniques have included the use of biotechnology and engineering technologies at the cellular level in recent years. Microorganisms, both autotrophic and heterotrophic, can be genetically modified to boost their CO2 sequestration ability, notably by increasing microbial CO2 fixing and decreasing CO2 release. Due to the presence of a complete CO2-fixing pathway and the ability to transfer energy from sunlight and inorganic compounds into cellular metabolites, autotrophic bacteria have evolved to subsist only on CO2. Heterotrophic microbes, on the other hand, rely on organic substances to thrive [42]. Therefore, the autotrophs could be engineered to improve the efficiency of their CO_2-fixing pathway, energy-harvesting systems and to regulate their cell resources, whereas the heterotrophs could be engineered to improve their carboxylation reactions in the metabolic pathways, to establish non-native CO_2-fixing bypass and ultimately to engineer them into autotrophs (**Table 1**).

Microorganisms	Targets	Strategies	References
Autotrophs	Improve the efficiency of the CO_2-fixing pathway	(1) Regulate the expression of CO_2-fixing pathway enzymes; (2) Improve the catalytic properties of carboxylases; (3) Create synthetic CO_2-fixing pathways.	[43, 44]
	Developing and optimizing energy harvesting systems	(1) Optimize natural photosystems; (2) Create artificial photosystems; (3) Develop electricity utilizing systems.	[45, 46]
	Regulating cell resources	(1) Enhance the product synthesis pathway; (2) Engineer transcription factors; (3) Provide organic carbon resources.	[47, 48]
Heterotrophs	Improve carboxylation reactions in metabolic pathways	(1) Augment the activity of carboxylases; (2) Increase intracellular CO_2 availability.	[49, 50]
	Establish non-native CO_2-fixing bypass	(1) Establish autotrophs transferred non-native CO_2-fixing bypass; (2) Create artificial pathways.	[44, 51]
	Engineer heterotrophs into autotrophs	(1) Install complete CO_2-fixing pathways; (2) Equip energy harvesting systems.	[46, 52]

Table 1.
Selected strategies to improve microbial CO_2 sequestration.

Modifications of both autotrophic and heterotrophic microorganisms to increase their efficiency in CO_2 sequestration *via* genetic engineering approaches are highly promising strategies for mitigating climate change. Lower costs of production and the naturally rapid growth rate of soil microorganisms should accelerate its adoption as a reliable CO_2 sequestration strategy. Furthermore, microbial CO_2 sequestration can be applied directly to complement agricultural activities on the same land compared with conventional CO_2 sequestration technologies that require purpose-built infrastructures that compete for land resources [53].

4. Conclusion

Soil microbiome activity has a huge implication on soil ecosystem dynamics, generally by promoting soil fertility and plant productivity. Soil is also a storage for carbon bulk either as SOM or SOC in terrestrial ecosystems. Carbon storage is the result of symbiotic interactions between plants and microbes in soils, through dynamic ecological processes of photosynthesis, decomposition, and soil respiration. The interaction and carbon sequestration are complicated to be measured precisely. Nonetheless, various research in recent years has clarified that human activities and climate change have had a significant impact on the soil's ecosystem, thus necessitating effective carbon balancing measures. As the shift toward sustainable agriculture is strengthening, the carbon footprint is one point of interest to benchmark the level of sustainability in agriculture activities. Moving forward, many techniques for carbon balancing and mitigation in soils and plant dynamic systems can be used, both at the macroscopical and microscopical levels of soil management.

Author details

Mohd N.H. Sarjuni[1,2], Siti A.M. Dolit[1], Aidee K. Khamis[2], Nazrin Abd-Aziz[1,2], Nur R. Azman[1] and Umi A. Asli[1,2*]

1 Faculty of Engineering, School of Chemical and Energy Engineering, University of Technology Malaysia, Johor Bahru, Johor, Malaysia

2 Innovation Center in Agritechnology for Advanced Bioprocessing (ICA), University of Technology Malaysia Pagoh Campus, Pagoh, Johor, Malaysia

*Address all correspondence to: umi_aisyah@utm.my

References

[1] Xiong W, Jousset A, Li R, Delgado-Baquerizo M, Bahram M, Logares R, et al. A global overview of the trophic structure within microbiomes across ecosystems. Environment International. 2021;**151**:106438. DOI: 10.1016/j.envint.2021.106438

[2] Lederberg J, McCray AT. `Ome sweet `omics--a genealogical treasury of words. The Scientist. 2001;**15**:8-8

[3] Geisen S. The future of (soil) microbiome studies: Current limitations, integration, and perspectives. mSystems. 2021;**6**(4):e0061321. DOI: 10.1128/mSystems.00613-21

[4] Del Frari G, Ferreira RB. Microbial blends: Terminology overview and introduction of the neologism "skopobiota". Frontiers in Microbiology. 2021;**12**:659592. DOI: 10.3389/fmicb.2021.659592

[5] Abakumov E, Zverev A, Kichko A, Kimeklis A, Andronov E. Soil microbiome of different-aged stages of self-restoration of ecosystems on the mining heaps of limestone quarry (Elizavetino, Leningrad region). Open Agriculture. 2021;**6**(1):57-66. DOI: 10.1515/opag-2020-0207

[6] Nilsson RH, Anslan S, Bahram M, Wurzbacher C, Baldrian P, Tedersoo L. Mycobiome diversity: High-throughput sequencing and identification of fungi. Nature Reviews Microbiology. 2019;**17**(2):95-109. DOI: 10.1038/s41579-018-0116-y

[7] Too CC, Keller A, Sickel W, Lee SM, Yule CM. Microbial community structure in a Malaysian tropical peat swamp forest: The influence of tree species and depth. Frontiers in Microbiology. 2018;**9**:2859. DOI: 10.3389/fmicb.2018.02859

[8] Gispert M, Emran M, Pardini G, Doni S, Ceccanti B. The impact of land management and abandonment on soil enzymatic activity, glomalin content and aggregate stability. Geoderma. 2013;**202-203**:51-61. DOI: 10.1016/j.geoderma.2013.03.012

[9] Malobane ME, Nciizah AD, Nyambo P, Mudau FN, Wakindiki IIC. Microbial biomass carbon and enzyme activities as influenced by tillage, crop rotation and residue management in a sweet sorghum cropping system in marginal soils of South Africa. Heliyon. 2020;**6**(11):e05513. DOI: 10.1016/j.heliyon.2020.e05513

[10] Trivedi P, Delgado-Baquerizo M, Anderson IC, Singh BK. Response of soil properties and microbial communities to agriculture: Implications for primary productivity and soil health indicators. Frontiers in Plant Science. 2016;7:990. DOI: 10.3389/fpls.2016.00990

[11] Wang X, He T, Gen S, Zhang X-Q, Wang X, Jiang D, et al. Soil properties and agricultural practices shape microbial communities in flooded and rainfed croplands. Applied Soil Ecology. 2020;**147**:103449. DOI: 10.1016/j.apsoil.2019.103449

[12] Guo S, Wu J, Coleman K, Zhu H, Li Y, Liu W. Soil organic carbon dynamics in a dryland cereal cropping system of the loess plateau under long-term nitrogen fertilizer applications. Plant and Soil. 2011;**353**(1-2):321-332. DOI: 10.1007/s11104-011-1034-1

[13] Holík L, Hlisnikovský L, Honzík R, Trögl J, Burdová H, Popelka J. Soil

microbial communities and enzyme activities after long-term application of inorganic and organic fertilizers at different depths of the soil profile. Sustainability. 2019;**11**(12):3251. DOI: 10.3390/su11123251

[14] Hassink J, Whitmore AP. Preservation of plant residues in soils differing in unfilled protective capacity. Soil Science Society of America Journal. 1996;**60**(2):487-491

[15] Paustian K, Six J, Elliott ET, Hunt HW. Management options for reducing CO2 emissions from agricultural soils. Biogeochemistry. 2000;**48**:147-163

[16] Karmakar R, Das I, Dutta D, Rakshit A. Potential effects of climate change on soil properties: A review. Science International. 2016;**4**:51-73. DOI: 10.17311/sciintl.2016.51.73

[17] Huang J, Zhang C, Cheng D, Hu B, Zhang P, Wang Z, et al. Soil organic carbon mineralization in relation to microbial dynamics in subtropical red soils dominated by differently sized aggregates. Open Chemistry. 2019;**17**(1):381-391. DOI: 10.1515/chem-2019-0051

[18] Santos JA, Fraga H, Malheiro AC, Moutinho-Pereira J, Dinis L-T, Correia C, et al. A review of the potential climate change impacts and adaptation options for European viticulture. Applied Sciences. 2020;**10**(9):3092. DOI: 10.3390/app10093092

[19] Lal R. Fate of soil carbon transported by erosional processes. Applied Sciences. 2022;**12**(1):48. DOI: 10.3390/app12010048

[20] Huang J, Li Y, Fu C, Chen F, Fu Q, Dai A, et al. Dryland climate change: Recent progress and challenges. Reviews of Geophysics. 2017;**55**(3):719-778. DOI: 10.1002/2016RG000550

[21] Zhao F, Wu Y, Hui J, Sivakumar B, Meng X, Liu S. Projected soil organic carbon loss in response to climate warming and soil water content in a loess watershed. Carbon Balance and Management. 2021;**16**(1):24. DOI: 10.1186/s13021-021-00187-2

[22] Johnston ASA, Sibly RM. The influence of soil communities on the temperature sensitivity of soil respiration. Nature Ecology & Evolution. 2018;**2**(10):1597-1602. DOI: 10.1038/s41559-018-0648-6

[23] Zhao Y, Ding Y, Hou X, Li FY, Han W, Yun X. Effects of temperature and grazing on soil organic carbon storage in grasslands along the Eurasian steppe eastern transect. PLoS One. 2017;**12**(10):e0186980. DOI: 10.1371/journal.pone.0186980

[24] Manik SMN, Pengilley G, Dean G, Field B, Shabala S, Zhou M. Soil and crop management practices to minimize the impact of waterlogging on crop productivity. Frontiers in Plant Science. 2019;**10**:140. DOI: 10.3389/fpls.2019.00140

[25] Bertola M, Ferrarini A, Visioli G. Improvement of soil microbial diversity through sustainable agricultural practices and its evaluation by -omics approaches: A perspective for the environment, food quality and human safety. Microorganisms. 2021;**9**(7):1400. DOI: 10.3390/microorganisms9071400

[26] Bhattacharyya SS, Ros GH, Furtak K, Iqbal HMN, Parra-Saldívar R. Soil carbon sequestration – An interplay between soil microbial community and soil organic matter dynamics. Science of the Total Environment. 2022;**815**:152928. DOI: 10.1016/j.scitotenv.2022.152928

[27] Tiefenbacher A, Sandén T, Haslmayr H-P, Miloczki J, Wenzel W, Spiegel H. Optimizing carbon sequestration in croplands: A synthesis. Agronomy. 2021;**11**(5):882. DOI: 10.3390/agronomy11050882

[28] Zhang K, Maltais-Landry G, Liao H-L. How soil biota regulate C cycling and soil C pools in diversified crop rotations. Soil Biology and Biochemistry. 2021;**156**:108219. DOI: 10.1016/j.soilbio.2021.108219

[29] Ray R, Baum A, Rixen T, Gleixner G, Jana T. Exportation of dissolved (inorganic and organic) and particulate carbon from mangroves and its implication to the carbon budget in the Indian Sundarban. Science of the Total Environment. 2018;**621**:535-547. DOI: 10.1016/j.scitotenv.2017.11.225

[30] Liu Q, Xu H, Yi H. Impact of fertilizer on crop yield and C:N:P stoichiometry in arid and semi-arid soil. International Journal of Environmental Research and Public Health. 2021;**18**(8):4341. DOI: 10.3390/ijerph18084341

[31] Venter ZS, Jacobs K, Hawkins H-J. The impact of crop rotation on soil microbial diversity: A meta-analysis. Pedobiologia. 2016;**59**(4):215-223. DOI: 10.1016/j.pedobi.2016.04.001

[32] Wiesmeier M, Urbanski L, Hobley E, Lang B, von Lützow M, Marin-Spiotta E, et al. Soil organic carbon storage as a key function of soils - a review of drivers and indicators at various scales. Geoderma. 2019;**333**:149-162. DOI: 10.1016/j.geoderma.2018.07.026

[33] Davies J. The business case for soil. Nature. 2017;**543**(7645):309-311. DOI: 10.1038/543309a

[34] FAO. Recarbonization of Global Soils- A Tool to Support the Implementation of the Koronivia Joint Work on Agriculture. Food and Agriculture Organization of the United Nations. Rome, Italy: FAO; 2019. p. 12. Available from: https://www.fao.org/documents/card/en/c/ca6522en/

[35] Navarro-Pedreño J, Almendro-Candel MB, Zorpas AA. The increase of soil organic matter reduces global warming, myth or reality? Sci. 2021;**3**(1):18. DOI: 10.3390/sci3010018

[36] Krauss M, Ruser R, Müller T, Hansen S, Mäder P, Gattinger A. Impact of reduced tillage on greenhouse gas emissions and soil carbon stocks in an organic grass-clover ley - winter wheat cropping sequence. Agriculture, Ecosystems & Environment. 2017;**239**:324-333. DOI: 10.1016/j.agee.2017.01.029

[37] Brevik EC, Cerdà A, Mataix-Solera J, Pereg L, Quinton JN, Six J, et al. The interdisciplinary nature of SOIL. The Soil. 2015;**1**(1):117-129. DOI: 10.5194/soil-1-117-2015

[38] Li Q, Song X, Gu H, Gao F. Nitrogen deposition and management practices increase soil microbial biomass carbon but decrease diversity in Moso bamboo plantations. Scientific Reports. 2016;**6**(1):28235. DOI: 10.1038/srep28235

[39] Bastida F, Torres IF, Abadía J, Romero-Trigueros C, Ruiz-Navarro A, Alarcón JJ, et al. Comparing the impacts of drip irrigation by freshwater and reclaimed wastewater on the soil microbial community of two citrus species. Agricultural Water Management. 2018;**203**:53-62. DOI: 10.1016/j.agwat.2018.03.001

[40] Paustian K, Collier S, Baldock J, Burgess R, Creque J, DeLonge M, et al. Quantifying carbon for agricultural soil management: From the current status toward a global soil information system.

Carbon Management. 2019;**10**(6):567-587. DOI: 10.1080/17583004.2019.1633231

[41] Ramesh T, Bolan NS, Kirkham MB, Wijesekara H, Kanchikerimath M, Srinivasa Rao C, et al. Chapter one-soil organic carbon dynamics: Impact of land use changes and management practices: A review. In: Sparks DL, editor. Advances in Agronomy. Vol. 156. San Diego, CA: Academic Press; 2019. pp. 1-107. DOI: 10.1016/bs.agron.2019.02.001

[42] Hu G, Li Y, Ye C, Liu L, Chen X. Engineering microorganisms for enhanced CO2 sequestration. Trends in Biotechnology. 2019;**37**(5):532-547. DOI: 10.1016/j.tibtech.2018.10.008

[43] Behler J, Vijay D, Hess WR, Akhtar MK. CRISPR-based technologies for metabolic engineering in cyanobacteria. Trends in Biotechnology. 2018;**36**(10):996-1010. DOI: 10.1016/j.tibtech.2018.05.011

[44] Wu G, Yan Q, Jones JA, Tang YJ, Fong SS, Koffas MA. Metabolic burden: Cornerstones in synthetic biology and metabolic engineering applications. Trends in Biotechnology. 2016;**34**(8):652-664. DOI: 10.1016/j.tibtech.2016.02.010

[45] Sakimoto KK, Wong AB, Yang P. Self-photosensitization of nonphotosynthetic bacteria for solar-to-chemical production. Science. 2016;**351**(6268): 74-77. DOI: 10.1126/science.aad3317

[46] Shi L, Dong H, Reguera G, Beyenal H, Lu A, Liu J, et al. Extracellular electron transfer mechanisms between microorganisms and minerals. Nature Reviews Microbiology. 2016;**14**:651-662. DOI: 10.1038/nrmicro.2016.93

[47] Chen X, Hu G, Liu L. Hacking an algal transcription factor for lipid biosynthesis. Trends in Plant Science. 2018;**23**(3):181-184. DOI: 10.1016/j.tplants.2017.12.008

[48] Zhou J, Zhang F, Meng H, Zhang Y, Li Y. Introducing extra NADPH consumption ability significantly increases the photosynthetic efficiency and biomass production of cyanobacteria. Metabolic Engineering. 2016;**38**:217-227. DOI: 10.1016/j.ymben.2016.08.002

[49] Siegel JB, Smith AL, Poust S, Wargacki AJ, Bar-Even A, Louw C, et al. Computational protein design enables a novel one-carbon assimilation pathway. Proceedings of the National Academy of Sciences. 2015;**112**(12):3704-3709. DOI: 10.1073/pnas.1500545112

[50] Zhu LW, Zhang L, Wei LN, Li HM, Yuan ZP, Chen T, et al. Collaborative regulation of CO2 transport and fixation during succinate production in Escherichia coli. Scientific Reports. 2015;**5**(1):1-12. DOI: 10.1038/srep17321

[51] Clomburg JM, Crumbley AM, Gonzalez R. Industrial biomanufacturing: The future of chemical production. Science. 2017;**355**(6320):aag0804. DOI: 10.1126/science.aag0804

[52] Abernathy MH, He L, Tang YJ. Channeling in native microbial pathways: Implications and challenges for metabolic engineering. Biotechnology Advances. 2017;**35**(6):805-814. DOI: 10.1016/j.biotechadv.2017.06.004

[53] Venkata Mohan S, Modestra JA, Amulya K, Butti SK, Velvizhi G. A circular bioeconomy with biobased products from CO_2 sequestration. Trends in Biotechnology. 2016;**34**(6):506-519. DOI: 10.1016/j.tibtech.2016.02.012

Section 2

Chemistry and Geomechanics

Chapter 4

Soil Solution Chemistry in Different Land-Use Systems in the Northeast Brazilian Amazon

Juliana Feitosa Felizzola, Ricardo de Oliveira Figueiredo, Wenceslau Geraldes Teixeira and Bruno Carneiro

Abstract

For sustainable production systems, the nonuse of fire (where there is a greater loss of soil nutrients), would be a solution for the conservation of nutrients in the soil, with the use of management by cutting and grinding, introduction of AFSs (agroforestry systems), and maintenance of riparian vegetation. The concentrations of carbon and nutrients were evaluated in the soil solution in two small hydrographic basins in the municipality of Igarapé-Açu (Pará state) in the eastern Amazon region, Brazil. The evaluations were performed considering the biogeochemical cycling in six land-use classes—riparian forest, secondary forest (*capoeira*), pasture, slash-and-burn agriculture, chop-and-mulch agriculture, and an agroforestry system (AFS). The objective was to determine the effects of different land-use systems on the composition of the soil solution, aiming to recommend sustainable practices. The concentrations of nutrients were greater in the areas of slash-and-burn agriculture and pasture, indicating greater losses of these nutrients due to runoff and leaching. The loss of nitrate was highest in the slash-and-burn area, while the organic carbon and organic nitrogen losses were greatest in the riparian forest, then in the secondary forest and agroforestry areas.

Keywords: Agrosystems, nutrient cycling, slash-and-burn, riparian forest soil solution extractor

1. Introduction

The chemical composition of the soil solution reflects the soil–plant-organism interplay, whereby this solution transports nutrients and other elements that supply plants and the soil biota. The effects of different soil management practices on biogeochemical flows need to be monitored and understood [1]. To understand the flow of nutrients at the soil-water interface, studies of the composition of the soil solution consider the processes of entry and exit of ions in each land-use situation [2]. The chemical elements present in the soil solution from the decomposition of leaf litter and the composition of the soil depend on various factors, namely, microbial population, plant species, temperature, groundwater flow, and soil management practices.

Various processes affect the soil solution composition, especially the mineral phase of the soil, determined by the entry and exit of minerals between the solid phase (colloidal) and a liquid fraction (solution) of the chemical elements that interact with the terrestrial and aquatic components of a watershed [3]. The investigation of the spatial–temporal variations of the chemical components susceptible to leaching to the water table or subsurface flow to land areas with lower elevation, thus reaching the watercourse beds, relies on the collection of soil solution samples [4]. The microbasins of the two streams (locally called *igarapés*) studied here, Cumaru and São João, have drainage networks predominantly regulated by the volume of groundwater stored in the soil [5]. Depending on the land use and agricultural management, in periods of high rainfall, these watersheds are subject to peak flows, with a potential to increase the loss of nutrients by leaching into deep soil layers, a process also related to the capacity of the soil to retain these nutrients [6]. In natural conditions, the soils in the basins studied are chemically poor due to the material of origin, derived from nutrient-poor sediments with low cation-exchange capacity, besides the severe action of the climate (high temperature and rainfall), causing a high weathering rate, intense leaching of nutrients and water erosion [7, 8]. These factors result in the low availability of nutrients to crops. Despite this lack of nutrients, the soils in the upland areas (*terra firme*), mainly acrisols and ferralsols, have excellent physical properties, that is, they are deep, show high infiltration rates, and are permeable, penetrable, and well-drained [7].

The objective of this study was to evaluate the composition of the soil solution in two microbasins of the Maracanã River Basin, located in the municipality of Igarapé-Açu in the eastern Amazon, subject to different land-use systems. We aim to expand knowledge about the impacts of land-use systems on water resources, improve the soil and watershed management techniques, and thus contribute to the establishment of more sustainable production systems in the region, which is mainly populated by family farmers.

2. Materials and methods

2.1 Description and characteristics of the basins and areas studied

The study was conducted in two small paired basins (**Figure 1**), Cumaru (4,127 ha) and São João (2,518 ha), both of which are tributaries of the Maracanã River Basin, located in the municipality of Igarapé-Açu, Pará, Brazil. The two watersheds are located between latitudes from 1°12′00"S to 1°16′00"S and longitudes from 47° 32′00"W to 47°34′00"W. Igarapé-Açu is located in the Northeast Pará Mesoregion and Bragantina Microregion, about 120 km from the state capital, Belém. The main access routes are federal highway BR-316 and state highway PA-127. The two basins have easy access by dirt roads from the municipal seat, an important logistical factor for choosing them. These areas were selected because they have the predominant land-use systems in the region. They have highly permeable soils, predominantly by acrisols and ferralsols and gently rolling topography, cut by shallow channels (igarapés), with the presence of floodable marginal areas occupied by riparian forest stands (called igapós). The coverage mainly consists of secondary forested areas with different ages (capoeiras), small farm plots, and pastures with varying dimensions [9, 10]. Burning releases chemical compounds into the soil, making it temporarily fertile for crops. However, these compounds are rapidly lost to groundwater through leaching or carried away by runoff into the igarapés due to the physical characteristics

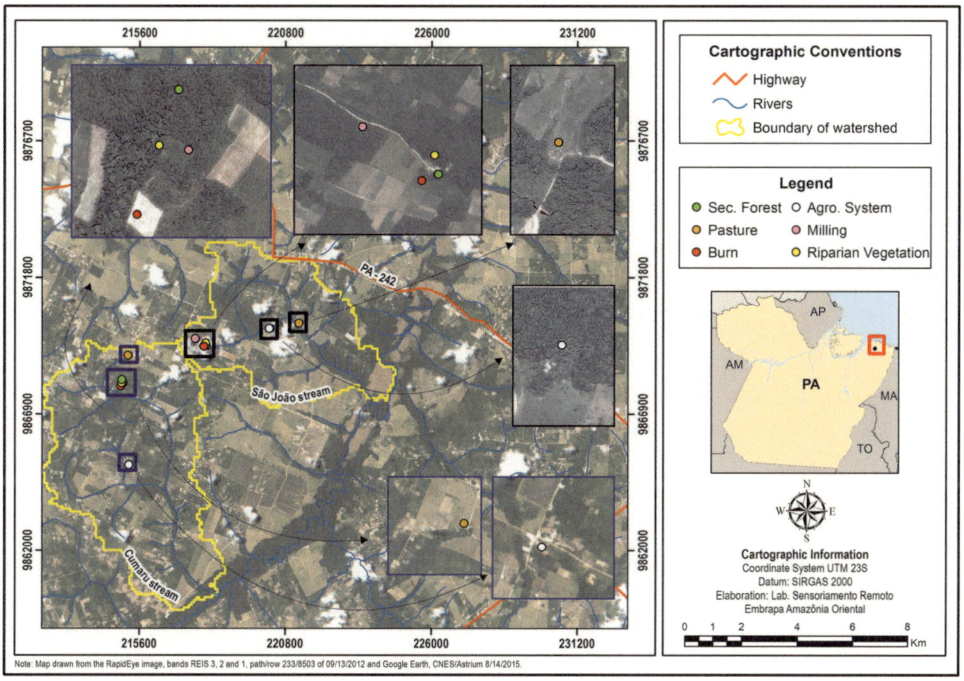

Figure 1.
The area was studied using soil solution extractors.

of the soil [11]. In a previous study, the pH of the soil in the Cumaru microbasin was found to vary from 4.8 to 4.9 at the depths studied, lower than the pH levels of the soils in the São João watershed, which ranged from 5.2 to 5.6 [7].

The predominant soils at the higher elevations of the Cumaru basin, where the relief is gently rolling, are classified as Xanthic Acrysol epicarenic, associated with a smaller proportion of Xanthic Ferralsol endoarenic, in addition to the occurrence of Petric Plinthosols. At lower elevations, the predominant soils are arenosols with hydromorphic characteristics, associated with small occurrences of Gleysols with indiscriminate texture in a narrow strip in ravines and other areas of frequent flooding of the drainage network. In the São João basin, the soils are very similar to those in the Cumaru basin, also with a predominant class of Xanthic Acrisolsepiarenic, associated with a smaller proportion of Xanthic Ferralsols endoarenic, along the soil with occurrences of lateritic concretionary horizons (Petric Plinthosols) at the higher elevations, with gently rolling terrain. At lower elevations, arenosols also prevail, with hydromorphic characteristics of sand or loamy sand texture in a very narrow strip near the streambed, widening slightly near the outlet of Igarapé São João into the Maracanã River [7]. They have acid pH, as mentioned, and low cation-exchange capacity (CEC), between 4.27 and 4.37, as described by Da Silva et al. [7].

The land-use characteristics of the areas studied are presented in **Table 1**. For each watershed, the location of the sampling points and the respective land use and agricultural management are identified, including the code assigned to each area. The two microbasins, although having similar land uses, differ regarding the intensity of these uses.

The riparian vegetation in Cumaru is near the beds of watercourses, which are classified as first and second order until reaching the Maracanã River, which is a

	Area Studied	Basin	Area Code	Latitude	Longitude
1	***Secondary Forest (<20 years)***	Cumaru	SFC secondary forest	47°33′40.48”W	1°11′30.64”S
2	***Slash-and-Burn Agriculture***	Cumaru	SBC slash-and-burn	47°33′ 41.44”W	1°11′37.92”S
3	***Chop-and-Mulch Agriculture***	Cumaru	CMC chop-and-mulch	47°33′ 39.88”W	1°11′34.59”S
4	***Pasture***	Cumaru	PC pasture	47°33′ 33.02”W	1°11′2.36”S
5	***AFS***	Cumaru	AFSC agroforestry system (rubber + cupuaçu palm)	47°33′ 32.09”W	1°13′10.03”S
6	***Riparian Vegetation***	Cumaru	RVC natural forest	47°33′ 40.73”W	1°11′34.53”S
7	***Secondary Forest (< 20 years)***	São João	SFSJ secondary forest	47°32′ 2.64”W	1°10′50.85”S
8	***Slash-and-Burn Agriculture***	São João	SBSJ slash-and-burn	47°32′ 5.28”W	1°10′50.85”S
9	***Chop-and-Mulch Agriculture***	São João	CMSJ chop and mulch	47°32′ 26.08”W	1°10′35.49”S
10	***Pasture***	São João	PSJ pasture	47°30′ 16.20”W	1°10′25.25”S
11	***AFS***	São João	AFSSJ agroforestry system (cupuaçu + peppertree)	47°30′ 49.76”W	1°10′31.50”S
12	***Riparian Vegetation***	São João	RVSJ natural forest	47°32′ 3.74”W	1°10′25.25”S

Table 1.
Location of the areas studied with different land-use classes.

third-order watercourse. The original vegetation of this basin was mainly equatorial forest, of which there are only a few remaining areas associated with the hygrophilous forest of alluvial plains around and along springs, streams, and rivers, where also exist hygrophilous floodplain forests. In the São João watershed, which also empties into the Maracanã River, the current prevailing forest regime is a secondary latifoliate forest in various development stages, resulting from clearance of the original equatorial forest, together with remnants of hygrophilous forest in alluvial plains along watercourses [7]. Riparian forest stands protect these igarapés and the fluvial water quality, and contribute to the groundwater stock, unlike what happens in areas subject to high surface runoff, mainly in unmanaged pastures, which in rainy periods suffer large losses of nutrients due to burning and the consequent absence of biomass from leaves and roots in the surface soil.

The pastures in both basins are unmanaged, without specific treatment. During the period studied, the pasture area in the Cumaru watershed contained about 1,200 head of cattle, while the pasture in the São João watershed contained only about 600 animals. For the slash-and-burn system, in both microbasins we prepared areas covering 0.5 ha in which cassava was planted. With respect to the secondary vegetation and agroforestry systems, these were already established in the two basins, so we demarcated areas of

0.5 ha. Finally, for the chop-and-mulch system, we also prepared areas of 0.5 ha, where we cut the secondary vegetation and left the biomass on the ground, composed of trunks, branches, and leaves. The ages of the secondary forest areas are given in **Table 1**.

2.2 Collection of samples and laboratory analyses

Soil solution extractors were installed six months before the monthly sampling campaigns, which occurred from March 2014 to April 2015. These extractors were installed in the Cumaru and São João watersheds in 12 areas measuring 0.5 ha in each basin, representing the land-use systems—riparian vegetation, secondary forest up to 20 years of recovering, slash-and-burn agriculture, chop-and-mulch agriculture, agroforestry system, and pasture. All told, we installed 96 extractors at depths of 30 and 60 cm (four at each depth) in the land plots in each use system.

The soil solution extractors consisted of porous capsules connected to amber glass collector jars (capacity of 250 mL) for analyses of DIC (dissolved inorganic carbon), DOC (dissolved organic carbon), and DON (dissolved organic nitrogen); and clear glass bottles (capacity of 1000 mL) for analyses of cations and anions—chloride (Cl^-), nitrate (NO_3^-), phosphate (PO_4^-), sulfate (SO_4^-), sodium (Na^+), ammonium (NH_4^+), potassium (K^+), magnesium (Mg^+), and calcium (Ca^+). The collector jars were washed after each sample collection with running water and 1% hydrochloric acid (HCl) solution, followed by Milli-Q water, and then dried and replaced.

We first measured the volume of water in the extractor jars. Then the samples were transported to the laboratory of Embrapa Amazônia Oriental in the city of Belém, for filtering and storage at ~4°C until conducting the analyses. For the analysis of DOC and DON, the samples were filtered through glass microfiber membranes (porosity of 0.7 μm), while for the other analyses, cellulose acetate membranes were used (porosity of 0.45 μm). To investigate the cations and anions, the samples were conserved in thymol ($C_{10}H_{14}O$) and submitted to ion chromatography with a chromatograph (Dionex DX120, USA).

The physical and chemical data of the soil samples from the region were obtained and described by Da Silva et al. [7].

2.3 Statistical analyses

We performed comparisons of the average concentrations in mg L^{-1} of the inorganic ions, cations, and anions, and organic and inorganic carbon dissolved in the soil solution samples from sites with different land uses over time. We also calculated correlations (Pearson correlation for parametric data and Spearman correlation for nonparametric data).

We used the Minitab 16.0 software to compute descriptive statistics and tests for normal distribution (Shapiro–Wilk test, $p > 0.05$) and comparisons of the means and variances. The parametric (normally distributed residues and homocesdatic variances) data were submitted to analysis of variance (ANOVA) for one factor (F test of statistical equality between means, Tukey test), while for interpretation of non-parametric data (not normally distributed), the Kruskal-Wallis test was used. After the mathematical transformation of the data, DON was the only parameter for which the values after this procedure had a normal distribution. The reciprocal transformation (1/Y) was used to stabilize the variance, in the sense of minimizing the effect of possibly very high values of Y. With this test, there was no evidence against the normality of the residuals because the points were all closely distributed along a straight line. In this case, one-way ANOVA was applied for a pairwise comparison of the means between land uses of the two microbasins. On the other hand, for the variables NO_3^-, NH_4^+, and DOC, the Kruskal-Wallis test was applied to compare the means of the nonparametric data. **Table 3** reports the mean values of the concentrations of the variables measured in relation to each land-use class.

2.4 Climate data

The climate in the region is Am according to the Köppen classification, hot and humid with an average annual rainfall of about 2,500 mm [12]. The climate data (**Figure 2**) were obtained from the records of the meteorological station of the Embrapa Eastern Amazon research unit (*Embrapa Amazônia Oriental*), which is located at latitude 01°11'S, longitude 47°35'W, and an altitude of 45 m, about 10 km from the sampling point. We collected samples in four climatic periods—rainy (March to June), transition 1- TR1 (July and August), dry (September to December), and transition 2- TR2 (January and February).

3. Results

This chemical variability revealed the important influence of land management and physical, chemical, and biological characteristics of the soil, along with seasonal climate, on the processes of entry and exit of nutrients in the watersheds studied.

3.1 Precipitation

Figure 2 presents the distribution of rainfall during the study period compared with the average of the previous five years. Of particular note is the drought that lasted from September 2014 to January 2015, when the temperature fluctuated between 26 and 28°C and the relative humidity averaged 85% in the year studied.

3.2 Variability of the volume of soil solution collected in the different areas studied

The greatest volumes collected by the soil solution extractors for evaluation of DIC, DOC, and DON were in the areas of riparian vegetation, AFSs, and secondary

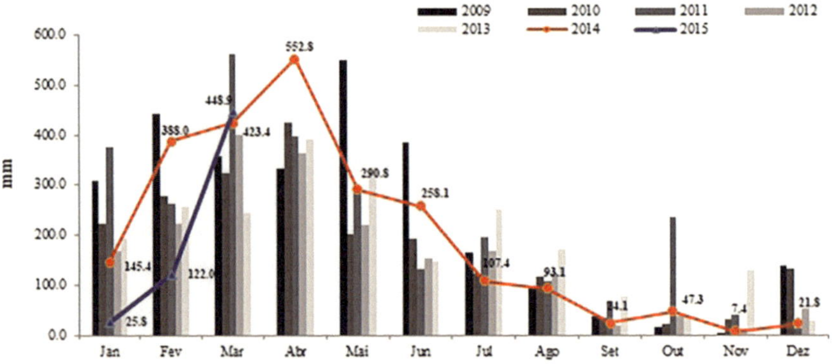

Figure 2.
Monthly precipitation from January 2009 to march 2015. Igarapé-Açu/Embrapa Amazônia oriental meteorological station.

SS	Study areas (depth 30 cm)											
Variables ($mg.L^{-1}$)	**PC**	**PSJ**	**SFC**	**SFSJ**	**AFSC**	**AFSSJ**	**RVC**	**RVSJ**	**CMC**	**CMSJ**	**SBC**	**SBSJ**
Cl^-	4.140 (1.47)	3.290 (1.89)	3.212 (0.63)	1.979 (0.65)	1.510 (0.33)	2.107 (0.44)	**7.88** (3.02)	1.747 (0.24)	1.580 (0.38)	2.89 (1.79)	3.94 (1.97)	1.242 (0.49)
NO_3^-	2.351 (0.97)	0.379 (0.11)	0.399 (0.16)	0.590 (0.25)	0.603 (0.14)	4.450 (1.74)	0.339 (012)	0.970 (0.26)	0.123 (0.10)	0.141 (*)	0.761 (0.38)	**8.227** (*)
PO_4^-	0.180 (*)	0.059 (0.02)	0.079 (0.02)	0.091 (0.04)	0.0705 (0.01)	0.051 (*)	0.100 (0.04)	* (*)	* (*)	0.879 (*)	**5.294** (*)	0.132 (0.07)
SO_4–	1.780 (1.05)	0.410 (0.12)	0.595 (0.23)	0.569 (0.125)	0.492 (0.26)	0.767 (0.20)	1.811 (0.55)	0.528 (0.07)	0.351 (0.11)	0.558 (0.38)	**1.98** (1.04)	0.857 (0.535)
Na^+	1.487 (0.73)	2.772 (0.78)	2.148 (0.40)	1.568 (0.325)	1.265 (0.16)	1.598 (0.43)	3.048 (0.84)	2.091 (0.45)	2.007 (0.83)	**3.24** (2.34)	2.46 (1.35)	0.789 (0.301)
NH_4^+	0.320 (0.11)	0.313 (0.15)	0.159 (0.07)	**0.353** (0.113)	0.1135 (0.04)	0.285 (0.01)	0.229 (0.08)	0.096 (0.03)	0.0461 (0.01)	* (*)	0.191 (0.09)	0.123 (0.001)
K^+	**4.390** (3.50)	3.037 (0.90)	2.250 (1.53)	0.730 (0.109)	0.433 (0.08)	0.692 (0.179)	0.722 (0.23)	1.415 (0.51)	0.729 (0.42)	1.64 (1.44)	6.67 (5.51)	2.677 (0.94)
Mg^{++}	0.675 (0.24)	0.801 (0.12)	0.413 (0.13)	0.386 (0.07)	0.295 (0.04)	0.487 (0.08)	0.443 (0.11)	**1.145** (0.46)	0.360 (0.12)	0.367 (0.28)	0.376 (0.24)	0.464 (0.07)
Ca^{++}	3.478 (0.66)	**3.854** (0.48)	1.754 (0.35)	1.942 (0.44)	1.731 (0.30)	2.6 (0.57)	0.954 (0.30)	1.155 (0.43)	1.351 (0.49)	1.545 (0.24)	2.049 (0.56)	2.61 (1.41)
DOC	8.020 (1.46)	27.20 (12.9)	9.900 (1.63)	19.32 (3.24)	3.305 (0.33)	8.31 (1.91)	10.21 (1.42)	16.41 (1.66)	10.86 (1.08)	8.59 (2.12)	15.32 (2.40)	**36.3** (24.5)
DON	1.147 (0.18)	2.032 (0.56)	1.151 (0.41)	1.459 (0.17)	0.916 (0.26)	0.998 (0.13)	0.812 (0.03)	1.291 (0.1)	0.916 (0.05)	1.083 (0.05)	**2.44** (1.48)	0.932 (0.12)
DIC	**1.404** (0.31)	1.136 (0.39)	0.365 (0.07)	0.406 (0.18)	0.898 (0.11)	0.938 (0.23)	0.224 (0.06)	0.291 (0.13)	0.574 (0.46)	0.833 (0.53)	0.835 (0.42)	0.639 (0.06)
SS	**Study Areas (depth 60 cm)**											
Ions ($mg.L^{-1}$)	PC	PSJ	SFC	SFSJ	AFSC	AFSSJ	RVC	RVSJ	CMC	CMSJ	SBC	SBSJ
Cl^-	5.000 (3.52)	1.793 (0.44)	1.702 (0.26)	4.26 (1.73)	7.08 (5.48)	2.006 (0.96)	5.929 (0.97)	2.773 (0.80)	**7.64** (5.04)	1.604 (0.63)	2.97 (1.38)	2.258 (0.45)

SS	Study areas (depth 30 cm)											
Variables (mg.L^{-1})	PC	PSJ	SFC	SFSJ	AFSC	AFSSJ	RVC	RVSJ	CMC	CMSJ	SBC	SBSJ
NO_3^-	2.190 (1.89)	0.500 (0.12)	0.925 (0.34)	0.546 (0.25)	2.03 (1.84)	2.93 (2.04)	0.367 (0.12)	3.79 (1.15)	2.57 (1.60)	1.733 (0.69)	**11.97** (9.66)	3.91 (3.47)
PO_4^-	* (*)	0.409 (*)	0.076 (0.008)	0.066 (*)	0.040 (0.005)	0.050 (*)	0.271 (0.21)	0.354 (0.24)	**34.8** (34.4)	0.788 (*)	1.066 (0.92)	0.259 (0.20)
SO_4–	3.940 (2.66)	0.235 (0.03)	0.716 (0.23)	0.373 (0.06)	4.63 (4.35)	0.629 (0.16)	1.443 (0.42)	0.988 (0.25)	**5.74** (4.51)	2.030 (0.88)	1.371 (0.5)	1.224 (0.34)
Na^+	4.100 (3.12)	1.454 (0.53)	1.605 (0.27)	2.477 (0.51)	3.89 (2.42)	1.381 (0.30)	2.754 (0.20)	3.78 (1.62)	**4.25** (1.41)	2.887 (0.83)	1.642 (0.30)	1.044 (0.33)
NH_4^+	0.272 (0.12)	0.205 (0.08)	0.154 (0.07)	0.1380 (0.03)	0.1010 (0.04)	0.1570 (0.06)	0.1935 (0.05)	1.117 (0.04)	0.568 (0.26)	0.035 (0.01)	**0.929** (0.21)	0.283 (0.13)
K^+	5.61 (4.87)	1.99 (0.93)	0.825 (0.33)	1.085 (0.32)	2.44 (2.21)	0.3880 (0.05)	1.018 (0.38)	1.154 (0.54)	**10.02** (9.16)	2.14 (1.85)	0.925 (0.28)	1.453 (0.32)
Mg^{++}	0.715 (0.36)	0.222 (0.031)	0.368 (0.07)	0.647 (0.23)	0.3185 (0.99)	0.327 (0.12)	0.3993 (0.07)	0.398 (0.20)	**0.722** (0.39)	0.699 (0.15)	0.50 (0.08)	0.533 (0.14)
Ca^{++}	3.398 (0.7)	**4.570** (1.24)	1.819 (0.49)	1.557 (0.51)	1.791 (0.41)	1.944 (0.70)	1.113 (0.33)	1.414 (0.38)	3.65 (1.23)	1.726 (0.26)	3.79 (1.61)	3.58 (1.18)
DOC	11.080 (1.57)	* (*)	6.220 (1.26)	12.6 (0.83)	3.408 (0.51)	6.08 (1.76)	2.231 (0.50)	**19.96** (3.83)	7.702 (0.73)	8.05 (1.38)	13.05 (1.33)	18.57 (2.21)
DON	1.215 (0.16)	1.036 (0.043)	0.8215 (0.07)	1.1257 (0.10)	0.6283 (0.04)	1.183 (0.34)	0.6551 (0.05)	1.329 (0.15)	0.6887 (0.03)	0.82 (0.06)	6.94 (3.76)	**1.574** (0.37)
DIC	1.120 (0.33)	1.561 (0.352)	0.4559 (0.09)	0.3396 (0.09)	0.678 (0.16)	0.398 (0.11)	0.352 (0.11)	2.26 (1.46)	**2.61** (2.18)	(*) (*)	0.554 (0.21)	1.341 (0.55)

Values in bold marked in yellow are the highest concentrations of the land-use type.

Table 2.
Concentrations of the chemical variables analyzed (mean and standard deviation) in the soil solutions of the areas studied in each basin in the different land-use classes at depths of 30 cm and 60 cm. PC = pasture Cumaru; PSJ = pasture São João; SFC = secondary Forest Cumaru; SFSJ = secondary Forest São João; AFSC = agroforestry system Cumaru; AFSSJ = agroforestry system São João; RVC = riparian vegetation Cumaru; RVSJ = riparian vegetation São João; CMC = chop-and-mulch agriculture Cumaru; CMSJ = chop-and-mulch agriculture São João; SBC = slash-and-burn agriculture Cumaru; SBSJ = slash-and-burn agriculture São João.

forest—a maximum of 198 mL and a minimum of 30 mL. In turn, for the analyses of the cations and anions, the greatest volumes were collected in the areas of riparian vegetation, AFSs, and chop-and-mulch agriculture, with a maximum volume of 850 mL and a minimum of 30 mL.

3.3 Spatial variability of nitrogen

The highest concentrations of nitrogen chemical species were found in the areas of slash-and-burn agriculture (**Tables 2** and **4**). The maximum concentration of NO_3^- in the SBC was 8.227 mg L^{-1} (at depth of 30 cm), while the concentration of NH_4^+ in this same area was 0.929 mg L^{-1} (at 60 cm), and DON was 2.44 mg L^{-1} (at 30 cm), as shown in **Table 3**. In turn, smaller concentrations of nitrogen occurred in the areas of chop-and-mulch agriculture and riparian vegetation. The concentrations of NO_3^-, considering the land-use systems, were higher than those of NH_4^+ and DON.

As can be observed in **Table 3**, there were significant differences according to the Kruskal-Wallis test in the mean values of the areas studied in the different land uses ($p \leq 0.05$) for DOC, NO_3^-, and NH_4^+. The results of the ANOVA (p-value = 0.000) provided sufficient evidence that the means were also different for DON, where $\alpha = 0.05$. According to the Tukey test, the AFS and riparian vegetation in Cumaru, as well as the pasture and riparian vegetation in São João, were different in relation to the other areas.

The boxplots in **Figure 3** present the variation of DON between the areas studied, where some discrepant data can be observed. In the Cumaru basin, it was highest in the areas of agroforestry system (AFSC) and riparian vegetation (RVC), while in São João the lowest values of DON were in the areas with secondary forest (SFSJ), pasture

			Kruskal-Wallis Test				One Way ANOVA-Tukey Test	
Land Use	DOC (p = 0.000)		NH_4 (p = 0.018)		NO_3^- (p = 0.004)		1/DON (p = 0.00)	
	Average Rank	Z	Average Rank	Z	Average Rank	Z	Mean	Grouping
SFC	118.7	−1.16	57.5	−1.13	61.8	−0.42	1.2943	A B
SFSJ	189.7	3.5	81.1	1.1	57.8	−0.67	0.9206	B C
PC	141.6	0.26	82.9	1.16	72.7	0.53	0.9598	B C
PSJ	194.5	2.8	69.1	−0.04	52.3	−1.01	0.8598	C
SBC	187.6	3.06	102.5	2.86	90.3	1.81	0.894	B C
SBSJ	192.5	3.6	74.7	0.32	73.4	0.49	1.1291	B C
AFSC	48.9	−5.98	49.2	−1.92	54.8	−1.31	1.5886	A
AFSSJ	95.1	−2.66	74.5	0.46	89.7	2.4	1.2731	A B
CMC	143.6	0.45	78.1	0.63	51.6	−1.06	1.2805	A B
CMSJ	126	−0.37	15.8	−2.35	77.3	0.71	1.1695	A B C
RVC	75	−5.53	68.3	−0.15	41.2	−3.31	1.5226	A
RVSJ	208.7	4.44	50.3	−1.66	87.4	2.41	0.8336	C

Table 3.
Statistical tests of the variables DOC, DON, NO_3^-, and NH_4^+.

Land use/Variable ($mg.L^{-1}$)	NO_3^-	SO42-	NH_4^+	K^+	CID	Mg2+	Na^+	COD	Cl^-	Ca^+	PO_4^-	NOD
PC	X	X	X	X	X	X				X		
PSJ			X	X	X	X	X			X		
SFC						X						
SFSJ			X	X		X				X		
AFSC		X			X	X	X					
AFSSJ	X					X						
RVC		X			X	X	X		X			
RVSJ	X		X			X	X					
CMC						X					X	
CMSJ						X						
SBC	X	X	X	X						X	X	X
SBSJ	X		X					X				

Table 4.
Presence of nutrients with highest concentrations (indicated by X) in the soil solution extracts according to land-use systems.

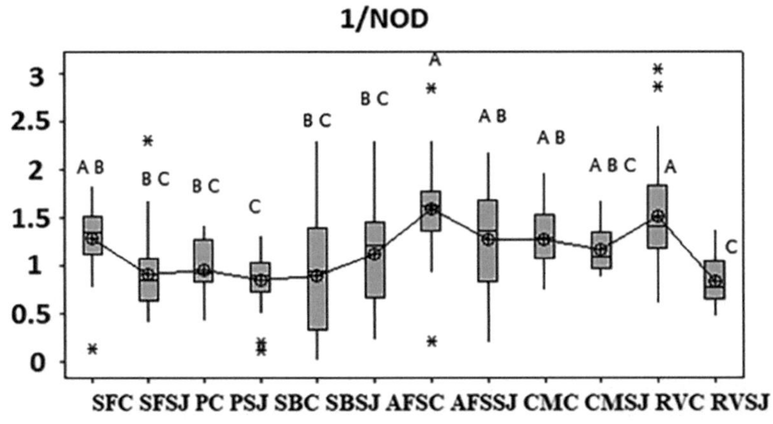

Figure 3.
Boxplots of the concentrations of dissolved organic nitrogen (DON) in the soil solution in the different areas studied after mathematical transformation (1/DON).

(PSJ), and agroforestry system (AFSSJ) in relation to the areas in Cumaru with the same land use.

3.4 Variability of DOC and DON in areas with different land-use systems

At a depth of 30 cm, there were higher concentrations of DOC in all land-use classes, but with slash-and-burn agriculture standing out more specifically in the

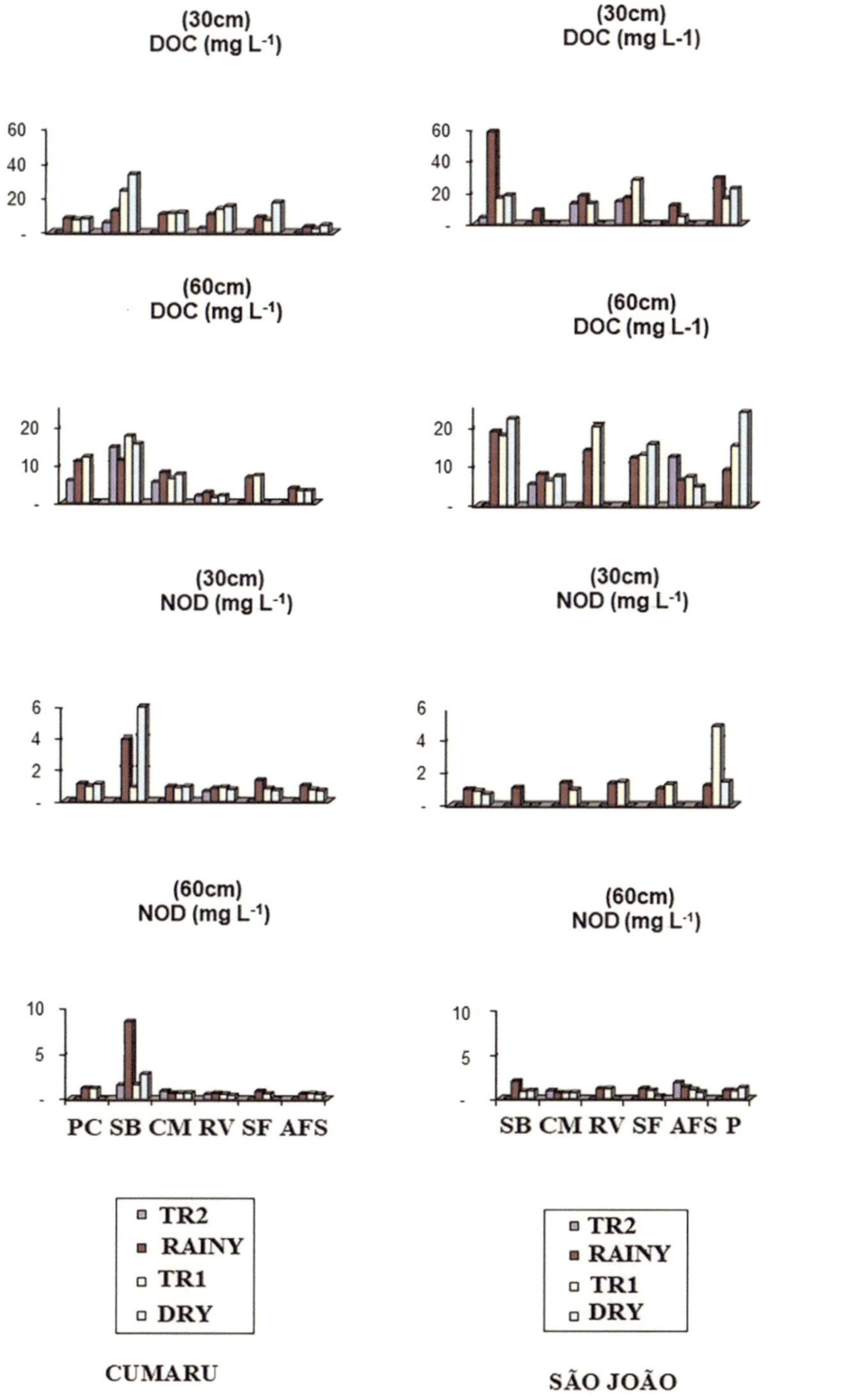

Figure 4.
Temporal and spatial variation of concentrations of dissolved organic carbon and nitrogen (in mg L^{-1}) in soil solution samples collected at depths of 30 and 60 cm in the sampling areas with different land-use classes in the watersheds studied.

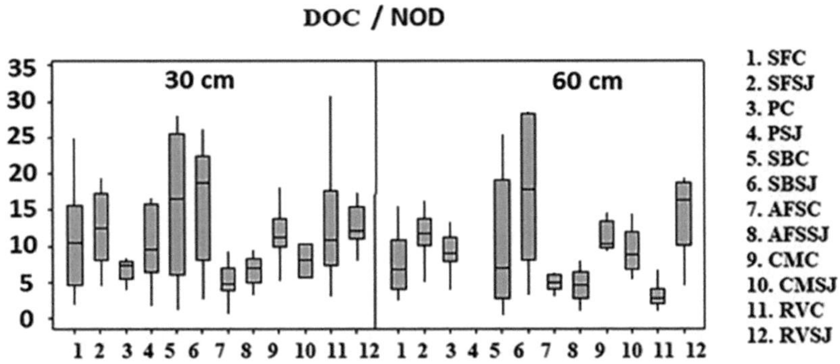

Figure 5.
The ratio between dissolved organic carbon and nitrogen in the agrosystems in the Cumaru and São João watersheds at depths of 30 cm and 60 cm.

periods of drought and transition from rainy to dry, with values above 20 mg L^{-1}. In turn, the riparian vegetation was in the second place regarding the magnitude of the concentrations of DOC. At a depth of 60 cm, slash-and-burn agriculture presented values greater than 10 mg L^{-1} of DOC in all climate periods (**Figure 4**).

In the pasture areas, the concentrations of DOC and DON were greater than 5 mg L^{-1} at both depths in the two transition periods and the rainy period, with no significant differences between depths. The DOC concentrations were lowest in the agroforestry system areas at both depths. With respect to climate, the DOC concentrations were greater in the dry period in all six land-use systems studied.

The DOC/DON ratio, in turn, was highest in the slash-and-burn agriculture areas and lowest in the riparian vegetation, AFS, and secondary forest areas (**Figure 5**).

Land Use	PC	PSJ	SFC	SFSJ	AFSC	AFSSJ	RVC	RVSJ	CMC	CMSJ	SBC	SBSJ
Anions mg.L^{-1}												
Cl^-												
30 cm	4.14	3.29	3.21	1.98	1.51	2.11	7.88	1.75	1.58	2.89	3.94	1.24
60 cm	5.00	1.79	1.70	4.26	7.08	2.01	5.93	2.77	7.64	1.60	2.97	2.26
NO_3^-												
30 cm	2.35	0.38	0.40	0.59	0.60	4.45	0.34	0.97	0.12	0.14	0.76	8.23
60 cm	2.19	0.50	0.93	0.55	2.03	2.93	0.37	3.79	2.57	1.73	11.97	3.91
PO_4^-												
30 cm	0.18	0.06	0.08	0.09	0.07	0.05	0.10	*	*	0.88	5.29	0.13
60 cm	*	0.41	0.08	0.07	0.04	0.05	0.27	0.35	34.80	0.79	1.07	0.26
SO_4–												
30 cm	1.78	0.41	0.60	0.57	0.49	0.77	1.81	0.53	0.35	0.56	1.98	0.86
60 cm	3.94	0.24	0.72	0.37	4.63	0.63	1.44	0.99	5.74	2.03	1.37	1.22

Land Use	PC	PSJ	SFC	SFSJ	AFSC	AFSSJ	RVC	RVSJ	CMC	CMSJ	SBC	SBSJ
Cations mg.L^{-1}												
Na^+												
30 cm	1.49	2.77	2.15	1.57	1.27	1.60	3.05	2.09	2.01	3.24	2.46	0.79
60 cm	4.10	1.45	1.61	2.48	3.89	1.38	2.75	3.78	4.25	2.89	1.64	1.04
NH_4^+												
30 cm	0.32	0.31	0.16	0.35	0.11	0.29	0.23	0.10	0.05		0.19	0.12
60 cm	0.27	0.21	0.15	0.14	0.10	0.16	0.19	1.12	0.57	0.04	0.93	0.28
K^+												
30 cm	4.39	3.04	2.25	0.73	0.43	0.69	0.72	1.42	0.73	1.64	6.67	2.68
60 cm	5.61	1.99	0.83	1.09	2.44	0.39	1.02	1.15	10.02	2.14	0.93	1.45
Mg^{++}												
30 cm	0.68	0.80	0.41	0.39	0.30	0.49	0.44	1.15	0.36	0.37	0.38	0.46
60 cm	0.72	0.22	0.37	0.65	0.32	0.33	0.40	0.40	0.72	0.70	0.50	0.53
Ca^{++}												
30 cm	3.48	3.85	1.75	1.94	1.73	2.60	0.95	1.16	1.35	1.55	2.05	2.61
60 cm	3.40	4.57	1.82	1.56	1.79	1.94	1.11	1.41	3.65	1.73	3.79	3.58
Organic and Inorganic Substances mg.L^{-1}												
DOC												
30 cm	8.02	27.20	9.90	19.32	3.31	8.31	10.21	16.41	10.86	8.59	15.32	36.30
60 cm	11.08	*	6.22	12.60	3.41	6.08	2.23	19.96	7.70	8.05	13.05	18.57
NOD												
30 cm	1.15	2.03	1.15	1.46	0.92	1.00	0.81	1.29	0.92	1.08	2.44	0.93
60 cm	1.22	1.04	0.82	1.13	0.63	1.18	0.66	1.33	0.69	0.82	6.94	1.57
CID												
30 cm	1.40	1.14	0.37	0.41	0.90	0.94	0.22	0.29	0.57	0.83	0.84	0.64
60 cm	1.12	1.56	0.46	0.34	0.68	0.40	0.35	2.26	2.61	*	0.55	1.34

Table 5.
Average concentrations of nutrients present in the soil solution in the Cumaru and São João basins in areas with slash-and-burn agriculture, riparian vegetation, secondary forest, chop-and-mulch agriculture, pasture, and agroforestry system.

3.5 Variability in the concentration of ions in areas with different land-use systems

The concentrations of the ions studied presented heterogeneous results in the different land-use systems, with the highest concentration found in the slash-and-burn agriculture, pasture, and riparian vegetation areas, as shown in **Table 5**.

With relation to depth, the ions concentration was variable between 30 and 60 cm. Pasture and chop-and-mulch stood out for having many high values at 60 cm (**Table 5**).

There were low correlations of DOC and DIC with Ca, R^2 = 2.1% and 15%, respectively (Spearman correlation, $p < 0.05$) (**Figure 6**).

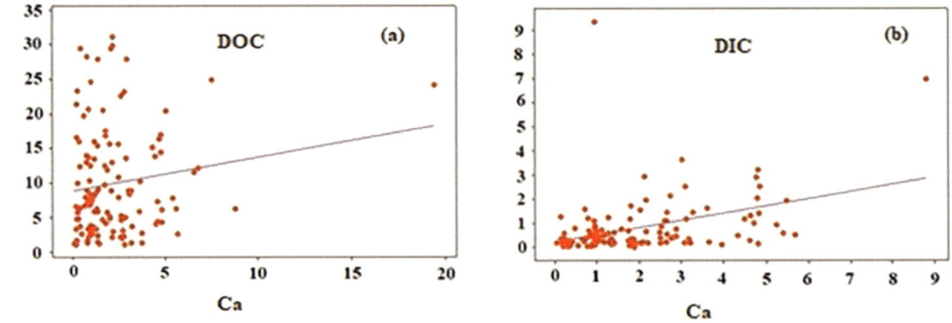

Figure 6.
Spearman correlations between dissolved organic carbon (DOC) versus calcium (a) and dissolved inorganic carbon (DIC) versus calcium (b).

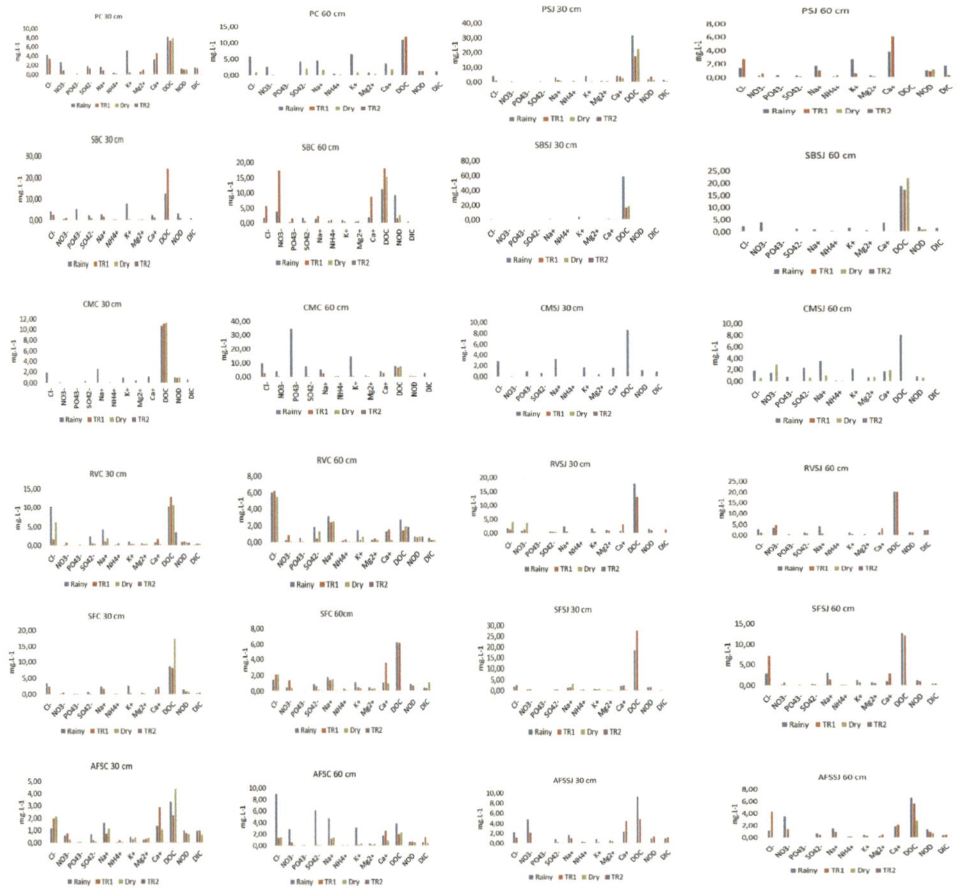

Figure 7.
Temporal distribution of the chemical constituents of the soil solution samples collected at depths of 30 and 60 cm in the 12 sampled areas.

3.6 Temporal variability of chemical compounds in areas with different land uses

With respect to the temporal variability, the highest concentrations of ions occurred in the rainy and transition to dry periods in all six land-use systems (**Figure** 7).

RV: The concentrations of DOC and DON were highest in the riparian vegetation area in both microbasins. This was the only land-use class for which we managed to obtain soil solution samples in all four climate periods. In the Cumaru microbasin, we detected all ions in the rainy and TR1 periods. The highest concentrations of Cl^- (approximately 10 mg L^{-1}) were found at 30 cm depth. In the São João microbasin, the highest concentrations of DOC, Cl^-, and NO_3^- occurred in the rainy, TR1, and dry periods.

SF: In the SFC, the concentrations of DOC and DON were highest in the rainy, TR1, and dry periods at both depths, while the inorganic ions stood out only in the rainy and TR1 periods. In SFSJ, the concentrations of these compounds were highest at depth of 30 cm, except for the phosphate levels in the secondary forest areas, which were below the limit of detection.

P: In the pastures, the concentrations of DOC and DON were highest at the 30 cm depth, principally in the São João microbasin. SO_4^{2-}, K^+, and Ca^{2+} were the predominant ions in the land-use areas in both watersheds.

SB: In the slash-and-burn agriculture areas, the concentrations of DOC were higher than 10 mg L^{-1}, with the greatest values finding at a depth of 60 cm in the rainy period and the transition from rainy to dry period in Cumaru. At the 30 cm depth, in the rainy period, the highest levels were measured of $NO_3^- \approx 16$ mg L^{-1}, $PO_4^- \approx 5$ mg L^{-1}, and $K^+ \approx 8$ mg L^{-1}). The same pattern occurred for these parameters in the São João microbasin, but with lower values in the rainy, TR1, and dry periods.

CM: In the chop-and-mulch system, phosphate stood out with the highest concentration (≈ 35 mg L^{-1}) at depth of 60 cm in the rainy period in the Cumaru microbasin. The other ions in this system had lower concentrations in the rainy, TR1, and dry periods, with values below 15 mg L^{-1} in both basins at the two depths.

AFS: In the agroforestry system, the ions had low concentrations, except for DOC, which had the highest concentrations in Cumaru at a depth of 60 cm and in São João at 30 cm, with approximate values of 9 mg L^{-1}. DOC and DON in the AFS sampling areas presented the highest concentrations in the dry period in Cumaru and the rainy period in São João.

4. Discussion

4.1 Temporal variability of C and N in the areas with different land uses

The concentrations of dissolved organic carbon at a depth of 30 cm were higher than at 60 cm; however, Marques et al. [13] did not observe differences between those two depths in a study in the Central Amazon near the city of Manaus, Brazil.

We found the concentrations of DOC in the slash-and-burn agriculture areas, mainly in the Cumaru microbasin, to be highest in the transition from the rainy to dry period (16.23–24.02 mg L^{-1}) and also in the dry season (15.39–33.48 mg L^{-1}). These are the periods when farmers most often practice burning.

In the São João microbasin, the maximum levels of DOC were also found in the slash-and-burn agriculture area at depth of 30 cm (57.86 mg L^{-1}). In this area, the soils are predominantly sandy rather than clayey, resulting in little retention of DOC and consequently a substantial presence in the soil solution, as also observed by Sommer [11] and Wickel [5].

The secondary forest and riparian vegetation areas presented the highest concentrations of dissolved organic carbon and nitrogen, probably originating from the decomposition of leaves and roots and the activity of the microbial biomass in the soil, among other factors, such as higher temperature and flow of the soil solution in the transition period TR1 (DOC: 1.37–19.98 mg L^{-1}; DON: 0.57–1.14 mg L^{-1}) and rainy period (DOC: 2.70–17.65 mg L^{-1}; DON: 0.67–1.39 mg L^{-1}). This is similar to the findings of Marques et al. [13], $\approx$ 10 mg L^{-1} for DOC. The spatial variations, that is, according to land use, can be observed in the DOC/DON ratio, where low ratios indicate higher concentrations of DON, which stand out in the agroforestry systems. The flows of DOC and DON are connected by belonging to the same organic matter chain in the soil. Part of the DOC is used in the soil as substrate by microorganisms, causing an increase in the mineralization of DON and consequently the nitrification process. The concentration of DOC in the soil solution can also affect the speed of the denitrification process and the concentration of DON, as well as the flows of nutrients and metalloids in the soil, also depending on the pH, redox potential, and cation-exchange capacity [2, 14].

Riparian vegetation, AFS, and secondary forest systems are known to retain nutrients in the soil, with only small losses due to leaching [15] because of the quality and quantity of residues produced by the plant cover. With respect to the nitrogen compounds, the concentrations of nitrate were higher than those of ammonium and dissolved organic nitrogen in the six systems (DON: 0.6–1.7 mg L^{-1}; NH_4^+: 0.035 mg L^{-1}; NO_3^-: 0.1–12.0 mg L^{-1}). This result was expected, since NO_3^- is more soluble than DON and is not retained in clay minerals such as NH_4^+ [2, 16], where the conversion of ammonium into nitrate happens rapidly. However, the highest concentration of NH_4^+ occurred in the riparian vegetation, where the nitrification process is slow. Alfaia [17] found the presence of NH_4^+ in floodplain areas rich in clays. Gruditz and Dalhammar [18] reported that the nitrification process (the transformation of ammonium into nitrite and nitrate) occurs at a pH of approximately 8.0, a level higher than in our study (in acidic soils). In the pastures, we observed the greatest variations of DOC and DON in the dry period at both depths. This response is likely due to the concentration of aggregates transported in the soil profile because in this period there is less water available [13].

4.2 Chemical variability in soil solution extracts based on different land uses

At both depths, ion exchange probably occurred, depending on the soil chemical composition and the presence of organic matter, in which the ions were carried through the unsaturated zones where plant roots are located, reaching the groundwater.

The different land uses influenced the nutrient cycling, depending on the management and complexity of the landscapes of the areas studied. In these areas, the soil composition is more sandy than clayey, except in the AFSs and pastures. In the unmanaged pasture and slash-and-burn agriculture areas, the adsorption of ions by the soil is hampered, mainly at the depth of 60 cm. This is characteristic of weathered soils with low cation-exchange capacity and high acidity. As described by other authors, the variations of pH and ionic strength are factors that influence the release of ions from the soil to the solution, and hence influence the processes of adsorption of cations and anions. In other words, acidic pH and low ion-exchange strength favored their release in the areas studied [5]. Besides this, the processes of nitrogen mineralization and subsequent oxidation (formation of nitrite and/or nitrate), along with the solubilization of aluminum, create conditions for the soil not to retain these ions [2]. The presence of iron and aluminum oxides in the soil can influence the electrochemical processes, resulting in increased exchange of anions and reduced exchange of cations between the soil and soil solution [16].

Competition exists between sulfate and nitrate ions, evidenced by the ionic strength and reduction of pH. This correlation was observed in the chop-and-mulch agriculture and AFS areas, with variations between these two parameters (**Table 2**). High concentrations of sulfate were observed in these areas, indicating the low competition of other anions such as phosphate and nitrate, since in these areas low concentrations of iron and aluminum predominated because the soils are sandier [7]. According to Borba et al. [16], sulfate ions are preferentially adsorbed by the soil in comparison with nitrate ions. The presence of sulfate in the soil solution in the AFS and chop-and-mulch agriculture systems can be attributed to soils with greater clay content than in areas of pasture and slash-and-burn agriculture, probably due to leaching of nutrients in soils impacted by external factors (physical–chemical destructuring) [15]. In this study, we observed a greater variation of Cl^- ions in the soil solution with lower retention, probably due to the competition for the adsorption of other negatively charged ions in the soil, such as NO_3^- and $SO4^-$.

With respect to cations, Ca^{2+}, Na^+, K^+, Mg^{2+}, and NH_4^+ had the highest concentrations in the chop-and-mulch agriculture and secondary forest systems, where the biomass left in the soil probably contributed to the incorporation of organic acids in the soil. This is an important factor for the enrichment of nutrients in the soil. With respect to the mechanism for the exchange of macronutrients and micronutrients, according to Borba et al. [16], this occurs mainly in shallower soil layers, where the greater presence of organic matter increases the cation-exchange capacity (CEC). According to Denich [19] and Cattanio [20], the level of organic material is affected by the composition of the leaf litter in secondary forest areas, in turn, determined by the plant species and their contribution of nutrients, as well as the pattern of decomposition and mesofauna in the soil. This same process probably occurred in the areas of riparian vegetation and agroforestry systems studied by us.

In the slash-and-burn agriculture areas, the concentration of nitrate in the soil solution was high, in some cases above the limits of potability (> 10 mg L^{-1}), indicating potential leaching into the soil solution, eventually reaching groundwater reservoirs. Williams et al. [21] conducted studies of the hydrochemical changes caused by land clearance through burning for agriculture in small plots in the central Amazon Basin on the northern side of the Solimões River and concluded that high leaching of nutrients occurs after slash-and-burn preparation, which gradually diminishes, mainly in the riparian zone. The authors also observed high concentrations and N and P in surface runoff, confirming the importance of riparian forests as nutrient buffers in the aquatic and terrestrial systems of the basin. In our study, in turn, we noted this same loss of micro-and macronutrients in the slash-and-burn agriculture and pasture areas.

The nitrogen in its inorganic form left after burning of vegetation is known to accelerate the decomposition of forest biomass [21], so that percolation of the solution through the soil profile in nitrate-rich soils occurs mainly in disturbed soils because this disturbance diminishes the nitrogen uptake of plants and enhance its availability. We found increased NH_4^+ in the soil solution to be common after burning and in areas with riparian vegetation of the Cumaru microbasin. This was probably due to microbiological processes, where ammonium is released from anoxic sediments to be oxidized into nitrate by nitrifying organisms, as well as caused by competition for N by plants [21].

Borba et al. [2] reported that the combination of dissolved inorganic carbon and Ca^+ ions, where carbon is converted into carbonates, caused an increase in soil pH. This linear correlation (**Figure 6**) between dissolved carbon and calcium can be related to the availability of this nutrient by the decomposition of biomass in agricultural systems, which, depending on the soil conditions (clay percent, acidity, and water availability) can increase or decrease the productivity of crops [22].

We observed higher concentrations of calcium, magnesium, potassium, and phosphate ions in the chop-and-mulch agriculture areas, probably resulting from the mineralization of the chopped plant residues, as reported by Figueiredo et al. [23] in a study of similar areas.

The concentration of ions in the soil increased rapidly after the burning of vegetation to plant crops, with various ionic compounds remaining for brief periods [6]. Changes in soil management are clearly related to loss and mobilization of nutrients affecting soil fertility, which is also affected by factors such as soil class, degree of disturbance, and effects of burning on the biological and physical aspects of the soil. The results of this study clearly indicated the importance of research regarding nutrient flows in areas with different soil management types in tropical areas, to evaluate the efficiency of different crops to absorb these nutrients without loss or excess of these compounds in the soil and water.

4.3 Temporal variability of nutrients in different land uses

We observed differences in the concentrations of carbon and nutrients in the soil solution in areas with different land uses and different climate periods (rainy, TR1, dry, and TR2) during the year in the two microbasins. To facilitate the discussion, we have divided the elements analyzed into three groups (**Figure** 7), which are Group 1 – dissolved organic carbon, dissolved organic nitrogen, and dissolved inorganic carbon; Group 2 – ammonium, potassium, and magnesium; and Group 3 – chloride, nitrate, phosphate, sulfate, sodium, and calcium.

In all land-use areas except pasture and AFSs, where the soils had a more clayey texture, the absence or presence of biomass (influenced by leaf concentration) along with rainfall affected the ions and organic compounds in the soil solution extracts, as also observed by Markewitz et al. [1], who reported greater nutrient flows in natural and secondary forests. In the riparian vegetation in the rainy season and the transition to the dry season (TR1), mainly in the latter case, there were greater movements of DOC, DON, and DIC at 30 cm depth compared to 60 cm depth. Additionally, chloride, sodium, calcium, and sulfate were present in higher concentrations than ammonium, potassium, and magnesium. Rainwater and its interaction with the canopy and litter affect the chemical composition of the soil solution, which is enhanced by the fact that the soil is sandy, causing little influence given its lesser capacity to retain nutrients. The greater presence of calcium and DIC in secondary forest areas can be related to the different compositions of trees (fewer species), probably due to the increased pH, favoring cation concentration in the soil. These factors were also observed by Markewitz [24] and Figueiredo [23], who described a substantial increase of cationic components in secondary vegetation. Despite the losses by leaching in this type of vegetation, the roots remain in the soil, where they form a protective network, hindering the passage of nutrients to watercourses [5].

The soils in the pastures at the two depths (30 and 60 cm) stood out for high concentrations of potassium and calcium in the rainy and TR1 periods. The mobility of these ions in the soil profile is intense after heavy rainfall, so their concentration in the soil solution increases. This is common in tropical soils in unmanaged pasture areas [25]. The soils in these areas, with low CEC, sparse organic matter, and high acidity, are prone to leaching due to high water percolation [26]. This process influenced the soil solution, mainly in the pasture area of the São João microbasin, where the DOC concentration reached 30 mg L^{-1} in the rainy season. The slash-and-burn areas of the two microbasins stood out with the loss of carbon from the soil solution via DOC, and of nitrogens via nitrate (NO_3^-) and DON, at the depth of 60 cm. This probably favored the movement of these compounds to the groundwater, as also observed by other authors [23, 24]. The amounts of DON and DOC lost to the soil solution in the process of burning vegetation and the decomposition of organic matter cause an increase in the flow of nitrate. These processes in the rainy season and transition to the dry season in the burned areas resulted in the loss of these compounds, which were intensely released from the soil surface by rainwater, rather than resulting from processes in the deeper soil levels, where they are transported to the groundwater.

In both microbasins, there were greater concentrations of phosphate at a depth of 60 cm in the chop-and-mulch agriculture areas in the rainy season, as well as of DOC with origin in the biomass on the soil surface, as also reported by Kato et al. [6]. According to Markewitz et al. [24], phosphorus can be converted to its inorganic state during the decomposition of organic matter. In our case, this nutrient in the soil came from the chopped branches and leaves. This process was also reported by Neill et al. [27] as the result of burning. This did not occur in this study, where it only happened in the chop-and-mulch area when the biomass supplied phosphorus to the surface soil and consequently to the soil solution. Kato et al. [6] noted a positive balance of nutrients, mainly phosphorus, through the cutting and mulching of secondary vegetation, which served as a source of organic matter to the soil. We also observed intense retention of water by the soil in this system, which hampered the collection of samples by the soil extractors. This was expected of the chop-and-mulch system as a factor of water conservation.

In the agroforestry systems, the concentrations of all ions were lower in the rainy, TR1, and dry seasons in relation to the other land uses. Ca^{2+}, Cl^-, Na^+, NO_3^-, and SO_4^- were present, along with DOC, DON, and DIC. The diversity of leaves from different plant species in the soil promotes the entry of nutrients because the presence of roots and microorganisms favors their absorption with less water availability [28]. Because of the presence of many arboreal species, studies of nutrient cycling in this complex composition are still scarce, but AFS models, implemented mainly by family farmers, have proven to be well-adapted, helping to improve local socioeconomic conditions [29].

5. Conclusions

- The presence of cations and anion were greater in the soil solution samples in the two transition periods (rainy to dry and vice versa).

- The cations were retained more in systems with a greater diversity of plants (riparian vegetation, secondary forest, and AFS).

- The sandier soils characteristic of these areas facilitated the leaching of nutrients into the soil solution, mainly with the practice of slash-and-burn agriculture.

- According to the data obtained, the systems in which there is more vegetation, such as secondary vegetation, riparian, and agroforestry systems, are the ones that retain nutrients in the soil, also preserving a greater amount of water in these systems. Unmanaged pasture and the use of burning for plantations have the risk of enhancing leaching rates by impoverishing the soil, so the use of crushed biomass in the soil agroforestry system and secondary vegetation can favor more sustainable management of the soil, in addition to the preservation of riparian forests.

Author details

Juliana Feitosa Felizzola[1*], Ricardo de Oliveira Figueiredo[2], Wenceslau Geraldes Teixeira[3] and Bruno Carneiro[4]

1 Laboratório de Análises de Sistemas Sustentáveis (LASS), Embrapa Amazônia Oriental (EMBRAPA), Travessa Dr. Enéas Pinheiro, Belém, Brazil

2 Departamento de Meio Ambiente, Embrapa Meio Ambiente (EMBRAPA), Jaguariúna, Brazil

3 Embrapa Solos, Rio de Janeiro, Brazil

4 Instituto Evandro Chagas (IEC), Seção Meio Ambiente, Embrapa Meio Ambiente (EMBRAPA), Ananindeua, Brazil

*Address all correspondence to: juliana.felizzola@embrapa.br

References

[1] Meurer EJ, Anghinoni I. A solução do solo. In: Meurer E, editor. Fundamentos da química do solo. 2nd ed. Vol. 2004. Porto Alegre: Genesis; 2004. pp. 101-129

[2] Borba RP, Ribeirinho VS, Camargo OA, Andrade CA, Kira CS, Coscione AR. Ion leaching and soil solution acidification in a vadose zone under soil treated with sewage sludge for agriculture. Chemosphere. 2018;**192**:81-89

[3] Pérez DV, Campos RC. Solução do solo: Importância e extração por centrifugação. Vol. 48. Rio de Janeiro: RJ - Embrapa solos, documentos; 2003. p. 36

[4] Conradie EC, Smith VR. Spatial variation in soil chemistry on a sub-Antartic Island. Open Journal of Soil Science. 2012;**2**(2):111-115

[5] Wickel B. Water and nutrient dynamics of a humid tropical agricultural watershed in Eastern Amazonia [thesis]. Bonn, Germany. Ecology and Development Series No. 21: University of Bonn; 2004. p. 135

[6] Kato Osvaldo, Kato Maria do Socorro A, de Carvalho Claudio José R, Figueiredo RO, Camarão AP, et al. "Uso de Agroflorestas no Manejo de Florestas Secundárias", In: da Gama-Rodrigues AC, de Barros NF, da Gama-Rodrigues EF, Freitas MSM, Viana AP, et al. Stemas Agroflorestais – Bases Científicas para o Desenvolvimento Sustentável. Campos dos Goytacazes: Universidade Estadual do Norte Fluminense; 2006. pp. 119-38

[7] Da Silva BNR, Silva LGT, Rodrigues TE, Gehard P. Solos das mesobacias hidrográficas dos igarapés São João e Cumaru, municípios de Marapanim e igarapé Açu. Congresso Brasileiro De Ciência Do Solo. Fortaleza: Embrapa Amazônia Oriental; 2009. p. 32

[8] Halden DA, França CF. Dinâmica do uso e ocupação do solo no município de Igarapé-Açu Pará, entre 1989 e 2008. vol. 8(9). Revista Perspectiva Geográfica Unioeste; 2013

[9] Corréa JM, Gerhard P, de Oliveira Figueiredo R. Ictiofauna de igarapés de pequenas bacias de drenagem em área agrícola do Nordeste Paraense. Vol. 7(2). Ambi-Agua, Taubaté: Amazônia Oriental; 2012. pp. 214-230. Available from: http://dx.doi.org/10.4136/ambiagua.739

[10] Gerhard P. Série Documentos: Embrapa Amazônia Oriental. Vol. n. 307. Belém: Definições de índices de integridade biótica; 2007

[11] Sommer R, Vlek PLG, Sá TD de A. Nutrient balance of shifting cultivation by burning or mulching in the Eastern Amazon – evidence for subsoil nutrient accumulation. Nutrient Cycling in Agroecosystem. 2004;**68**:257-271

[12] Araújo PN. Boletim agrometeorológico. In: Igarapé - Açu/por Nilza Araujo Pachêco e Therezinha Xavier Bastos. Belém, PA: Embrapa Amazônia Oriental; 2006. 30 p. il. 21cm. – (Embrapa Amazônia Oriental. Documentos, 296)

[13] Marques JDDEO, Luizão FJ, Teixeira WG, Ferreira SJF. Variações do carbono orgânico dissolvido e de atributos físicos do solo sob diferentes sistemas de uso da terra na Amazônia Central. Revista Brasileira de Ciência do Solo. 2012;**36**:611-622

[14] Bolan N, Kunhikrishan A, Thangarajan R, Kumpiene J, Park J, Makino T, et al. Remediation of heavy metal(loid)s contaminated soils – to mobilize or to immobilize? Journal of Hazardous Materials. 2014;**266**:141-166

[15] Figueiredo RO. Processos hidrológicos e biogeoquímicos em bacias hidrográficas sob usos agrícola e agroflorestal na Amazônia Brasileira. Brasília, DF: Embrapa Informação Tecnológica: Alternativa agroflorestal na Amazônia em transformação; 2009. pp. 477-500

[16] Borba RP, Camargo OA, Kira CS, Coscione AR. NO_2^- and NO_3^-leaching and solubilization of Al in variable charge soils treated with sewage sludge. Environmental Earth Science. 2015;**74**: 4625–4638. DOI: 10.1007/s12665-015-4428-1

[17] Alfaia SS. Caracterização e distribuição das formas de nitrogênio orgânico em três solos da Amazônia Central. Acta Amazonica. 2006;**36**(2): 135-140

[18] Grunditz C, Dalhammar G. Development of nitrification inhibition assays using pure cultures of Nitrosomonas and Nitrobacter. Water Research. 2001;**35**:433-440

[19] Denich M. Estudo da importância de uma vegetação secundária nova para o incremento da produtividade do sistema de produção na Amazônia Oriental Brasileira. Eschborn: Embrapa/cpatu-GTZ; 1991. p. 284

[20] Cattanio H. Soil mineralization dynamics as affected by pure and mixed application of leafy material from leguminous trees used in planted fallow in Brazil [thesis]. Göttingen, Germany: University of Gõttingen; 2002. p. 124

[21] Williams MR, Fisher TR, Melack JM. Solute dynamics in soli water and groundwater in a central Amazon catchment undergoing deforestation. Biogeochemostry. 1997;**38**:303-335

[22] Pias OHC, Tiecher T, Cherubin MR, Mazurana M, Bayer C. Crop yield responses to sulfur fertilization in Brazilian no-till soils: A systematic review. Revista Brasileira de Ciência do Solo 2019;**43**(1-21):e0180078

[23] Figueiredo RO, Markewitz D, Davidson EA, Schuler AE, Watrin OS, Silva PS. Land-use effects on the chemical atributes of low-order streams in the esatern Amazon. Journal of Geophysics Research. 2010;**115**:G04004

[24] Markewitz D, Davidson EA, Moutinho P, Nepstad D. Nutrient loss and redistribution after forest clearing on a highly weathered soil in Amazônia. Ecological Applications. 2004;**14**:S177-S199

[25] Markewitz D, Eric A, Davidson Ricardo d O, Victoria FRL, Krusche AV. Control of cations concentrations in stream waters by surface soil process in Amazonian watershed. Nature. 2001; **410**:802-805

[26] Hashimoto CV. Lixiviação de potássio em latossolo amarelo na Amazônia Central Brasileira, 153f. Faculdade de Engenharia: Tese de Mestrado – Universidade Federal do Rio de Janeiro; 2018

[27] Neill C, Deegan LA, Thomas SM, Cerri CC. Deforestation for pasture alters nitrogen and phosphorus in soil solution and streamwater of small Amazonian watersheds. Ecological Applications. 2001;**11**:1817-1828

[28] Lunz AMP, Franke IL. Princípios gerais e planejamento de sistemas agroflorestais. Rio Branco: Embrapa-CPAF/AC. 1998. 27p (Embrapa-CPAF/AC. Circular Técnica, 22)

[29] Vieira ICG, Toledo PM, Almeida A. Análise das modificações da paisagem da Região Bragantina, no Pará, Integrando diferentes escalas de tempo. Revista Ciência e Cultura. 2007;**59**:27-30

Chapter 5

CO_2 Injectivity in Deep Saline Formations: The Impact of Salt Precipitation and Fines Mobilization

Yen A. Sokama-Neuyam, Muhammad A.M. Yusof and Shadrack K. Owusu

Abstract

Climate change is now considered the greatest threat to global health and security. Greenhouse effect, which results in global warming, is considered the main driver of climate change. Carbon dioxide (CO_2) emission has been identified as the largest contributor to global warming. The Paris Agreement, which is the biggest international treaty on Climate Change, has an ambitious goal to reach Net Zero CO_2 emission by 2050. Carbon Capture, Utilization and Storage (CCUS) is the most promising approach in the portfolio of options to reduce CO_2 emission. A good geological CCUS facility must have a high storage potential and robust containment efficiency. Storage potential depends on the storage capacity and well injectivity. The major target geological facilities for CO_2 storage include deep saline reservoirs, depleted oil and gas reservoirs, Enhanced Oil Recovery (EOR) wells, and unmineable coal seams. Deep saline formations have the highest storage potential but challenging well injectivity. Mineral dissolution, salt precipitation, and fines mobilization are the main mechanisms responsible for CO_2 injectivity impairment in saline reservoirs. This chapter reviews literature spanning several decades of work on CO_2 injectivity impairment mechanisms especially in deep saline formations and their technical and economic impact on CCUS projects.

Keywords: CO_2 injectivity, mineral dissolution, salt precipitation, fines mobilization

1. Introduction

Since last decade, there has been a growing concern of the negative impacts of global climate change. In the last century, scientists believe that carbon dioxide (CO_2) emission is has been the main component responsible for approximately three-quarters of global greenhouse gas emission. A roadmap developed to combat climate change has outlined 10 scalable solutions clustered into categories of social transformative, governance improvement, market and regulation-based solutions, technological innovation and transformation, and lastly natural and ecosystem management. While some proposed mitigation techniques focus on reduction of CO_2 emission, Carbon Capture, Utilization and Storage (CCUS) technology can aim at achieving lower CO_2 amount in the atmosphere by capturing and storing the anthropogenic gas in a geological storage. Geological storage of CO_2 in depleted oil and gas reservoirs, deep saline reservoirs, unmineable coal seams or injected into active oil and gas reservoirs for Enhanced Oil Recovery (EOR) are currently a well-accepted method of storing CO_2. For economic reasons, CO_2 is being injected at the highest possible rates through limited number of wells. This could trigger injectivity-related issues due to complex interactions between CO_2, brine and rock initiated in the aquifer. This makes CO_2 injectivity not only a technical challenge but also an economic consideration.

This chapter presents a comprehensive discussion on how the various mechanisms contributed by the fluid-rock interactions during CO_2 sequestration affect CO_2 injectivity. The chapter begins by laying out the theoretical dimensions of CO_2 sequestration. This is followed by a brief overview of different CO_2 injectivity impairment mechanisms, focusing on the two main themes: salt precipitation and fines migration. The experimental findings from previous researchers have also been discussed and some findings remarks made. The insights gained from this study may be valuable to the rapidly expanding field of carbon sequestration.

2. Climate change, a global challenge

Climate Change is now considered the greatest threat to global health and security. Greenhouse effect, which propels global warming, has been identified as the main driver of Climate Change. The rising of global temperature is intricately linked with many other environmental concerns such as fragile ecosystem, melting glaciers, increasing sea level, acidification of sea water and increased flooding and droughts [1]. This climate challenge is also affecting the social community which can lead to immigration and conflicts over borders and natural resources such as water. More importantly, it could severely threaten food security that may affect about 3 billion of poor people in terms of access to food supply. These series of concerns are recognized as climate change and it is well accepted that to prevent its occurrence, greenhouse gas emission has to be reduced significantly over the twenty-first century [2]. The gases which are mainly responsible for the greenhouse effect include methane, carbon dioxide (CO_2), nitrous oxide, water vapor, and fluorinated gases.

CO_2 generated mainly from anthropogenic activities is the largest contributor to global warming. By 2020, the concentration of CO_2 in the atmosphere had risen to 48% above its pre-industrial level [3, 4]. An increase of 2°C above pre-industrial average temperature could induce serious negative impacts on the natural environment and human health. The Paris Agreement, which is the biggest international treaty on Climate Change is determined to limit global warming to about 1.5°C, compared to pre-industrial levels, with an ambitious goal to reach NetZero CO_2 emission by 2050 [5–7].

The anthropogenic activities with highest carbon footprint include the burning of fossil fuel for power generation and the production of materials. Power generation from fossil fuels is responsible for over 70% of the global CO_2 emissions [1]. Cumulatively, fossil fuel contributed about 84% of the World's primary energy consumption by 2019 and the world is expected to rely heavily on fossil fuels for its energy needs, at least within the short to medium term. Another major source of CO_2 emission is the production of materials with high carbon and energy footprint. The notable of such materials is Portland cement which is the main building material used in most countries in the world. Portland cement production is responsible for about 5% of global CO_2 emission [8]. About 2% of the total global energy consumption is used to produce Portland cement. Widespread use of energy efficient power generators, investing in renewable energy, Carbon Capture, Utilization and Storage (CCUS) remains the best options for reducing CO_2 emission from burning of fossil fuels and achieving a faster transition to green energy.

3. Geological storage of CO_2

CCUS is considered a viable option to reduce CO_2 emission, sustain exploration and production of fossil fuel for the short to medium term and eventually transition to a full green energy in the long term [9, 10]. Among the proposed CO_2 emission reduction strategies, CCUS provides the highest emission reduction potential [11]. Generally, CCUS involves the (1) capture of CO_2 from large industrial emission sources and direct air capture points (2) the transportation of the captured gas to utilization, conversion, or storage facilities and (3) the utilization of the gas as feedstock in industrial processes, conversion to other products or the injection of the gas into geological storage facilities. In terms of geological storage, the injected CO_2 may be stored in depleted oil and gas reservoirs, deep saline reservoirs, unmineable coal seams or injected into active oil and gas reservoirs for Enhanced Oil Recovery (EOR) [12, 13].

CCUS system, although simple in concept, would require significant investment of capital, new technology and time [14]. Besides, many current policies also need to be revised and new legal and regulations framework has to be introduced that require support from local authorities, governments and international bodies [15]. Investigation by the International Energy Agency (IEA) have shown that CCUS can contribute up to about 14% reduction in global greenhouse gas emissions required to limit global warming to 2°C by 2050 [16].

3.1 Storage in saline aquifers

Saline aquifer refers to a deep, large geological formation consisting permeable sedimentary or carbonate rock types that are saturated with formation water or brines, non-potable water, containing high concentration of dissolved salts [17]. It is buried under a layer of non- or low-permeability rocks that serve as a cap rock to prohibit the fluid flowing upwards to the surface. The saline aquifer can be located both onshore and offshore and normally found at depth greater (more than 800 metres) than aquifers that contain potable water [18]. Deep saline formations have enormous potential for CO_2 storage in terms of volumetric storage capacity [19, 20].

On a global scale, deep saline reservoirs have the capacity to hold between 20 and 500% of the projected CO_2 emissions by 2050 [9, 21, 22]. Thus, worldwide CO_2 storage potential of deep saline reservoirs ranges from 400 to 10,000 Gt CO_2. Deep saline aquifers, usually at depths between 700 and 1000 m, hold large quantities of high salinity formation brines [23].

Although the natural content of these reservoirs has no direct commercial value, the chemical composition of the formation brine makes them suitable for CO_2 mineralization. In deep saline aquifers, the injected CO_2 could be sequestered through hydrodynamic trapping where the gas is trapped beneath a caprock, residual trapping where the rock contains residual saturation of CO_2, solubility trapping where the gas dissolves in the formation brine and mineral trapping where CO_2 reacts with Ca, Fe or Mg to form stable carbonate precipitates [24, 25]. Lack of additional economic benefits except carbon tax incentives in some countries, makes CO_2 storage in saline aquifers less attractive to the oil and gas industry.

3.2 CO_2 enhanced oil recovery

CO_2-EOR is a tertiary oil recovery technique where CO_2 and usually other fluids such as water or brine is injected into the reservoir to achieve miscibility with the oil and recover residual oil. In addition to extraction of residual oil, the injected gas provides pressure support and could remain stored permanently after the recovery process. Under subsurface conditions, CO_2 mixes with oil above a certain minimum miscibility pressure (MMP), reducing the capillary effect that retain the oil in place [26–28]. There are four main underlying mechanisms of CO_2-EOR as outlined by Rojas and Ali [29] and Tunio et al. [30] which include (1) oil swelling; (2) reduction of oil viscosity; (3) reduction of oil and water density; and (4) extraction of oil components.

Alternative forms of CO_2-EOR have been developed over the past years, including continuous CO_2 injection, continuous CO_2 injection followed by water, water-alternating gas (WAG) and WAG followed by gas or water [31–33]. To improve sweep efficiency, carbonated water injection has also been used as a viable alternative [34, 35]. Other emerging injection schemes include CO_2 low salinity water alternating gas (CO_2-LSWAG) injection under miscible CO_2 displacement conditions [36–38]. Depleted oil and gas reservoirs are also attractive candidates for CO_2 storage due to the potential to reuse some of the production equipment and geological data collected over the producing life of the reservoirs to lower exploration cost and reduce the risk associated with CO_2 storage. It has been reported that depleted oil and gas reservoirs could hold about 45% of the projected CO_2 emissions by 2050 [9].

3.3 CO_2 injection into unconventional reservoirs

The two main unconventional formations where CO_2 injection is promising are coal seams and shale gas reservoirs. CO_2 enhanced coalbed methane recovery (CO_2-ECBM) has the potential to store large volumes of CO_2 in deep unmineable coal seams while improving the efficiency of coal bed methane recovery [39, 40]. The injected CO_2 displaces methane and remain sequestered in the coal seams as CO_2 is preferentially adsorbed onto coal seams, thus releasing the coal bed methane which can then be produced as free gas [41]. Based on the simple assumption that, for every CH_4 molecule, two molecules of CO_2 can be stored, IEA-GHG [42] estimated that about 220 GT of CO_2 could be stored in deep unmineable coal formations worldwide.

The potential to store CO_2 in organic-rich gas shales is attracting increasing interest, especially in countries that have extensive shale deposits [43–45]. Although still in early-stage research, CO_2 injection into organic-rich gas shales could provide dual benefits: an economic benefit from the incremental recovery of adsorbed methane, and an environmental benefit of secure CO_2 storage.

4. The prerequisites of CCUS

A viable candidate for CCUS must meet a threshold well injectivity required to inject large volumes of CO_2 at high injection rates through a minimum number of wells, adequate storage capacity to hold large volumes of CO_2 and robust containment to permanently isolate the sequestered gas from the environment [16]. Storage capacity and well injectivity defines the storage potential of a geological storage facility [11, 46, 47].

4.1 Storage capacity

Implementation of CCUS technology require accurate estimation of the pore space available in the reservoir rock to hold the injected CO_2 [48–51]. The storage capacity estimated can be of different levels of certainty and cost depending on the scale and resolution. The various CO_2 trapping mechanisms in deep saline aquifers, namely structural and stratigraphic trapping, residual gas trapping, solubility trapping, mineral trapping and hydrodynamic trapping, which occur at different times during the storage, must be considered in the estimation to obtain a representative estimate [51]. Other parameters that affect the storage capacity include in situ pressure, injectivity, temperature, permeability, and rock compressibility.

The volume of CO_2 that can be commercially sequestered in a reservoir within a specific period, using available technology, under current economic conditions, operating methods and governmental regulations has been termed the CO_2 storage reserve [11, 48]. The USDOE [52] has developed a simplified model to quantify the storage capacity of deep saline formations which is given by:

$$M_{CO_2} = V_A \phi_T \rho_{CO_2} E_s \tag{1}$$

In Eq. (1), M_{CO_2} is the mass of CO_2 that can be stored, V_A is the bulk volume of the aquifer, ϕ_T is the effective porosity of the aquifer, ρ_{CO_2} is the density of CO_2 at reservoir conditions and E_s is the storage efficiency. The storage efficiency expresses the degree of filling the reservoir [11], also defined as the ratio of the volume occupied by CO_2 to the total accessible pore volume of the reservoir [53]:

$$E_s = \frac{V_{CO_2}}{V_{pore}} \tag{2}$$

In Eq. (2), V_{CO_2} is the volume of injected CO_2 and V_{pore} is the accessible reservoir pore volume available for CO_2 storage. Eq. (1) and (2) can be coupled to estimate the volumetric CO_2 storage capacity of a given deep saline reservoir. CO_2 storage efficiency in deep saline formations depends on the reservoir rock properties (porosity, permeability, net to gross, thickness and area), the efficiency of water displacement by injected CO_2 and the degree of conformance of the aquifer [11].

Bachu et al. [54] have also proposed a model to estimate the theoretical CO_2 storage capacity of depleted oil and gas reservoirs, based on the assumption that the entire pore space originally occupied by hydrocarbons can be filled by CO_2 and that CO_2 can be injected until the reservoir pressure reaches the original pressure of the virgin reservoir. These assumptions can be valid if the reservoir is not in contact with an aquifer or already flooded during secondary and tertiary recovery. For practical purposes, an effective storage capacity could be defined to incorporate other important parameters such as displacement efficiency, gravity effects, residual oil and water saturation, reservoir heterogeneity, rock-fluid interactions, and formation damage.

4.2 Well injectivity

The injectivity of a reservoir measures the amount of CO_2 an injection well can receive without fracturing the formation [11]. Well injectivity can be expressed with an injectivity index, I, often defined as the ratio of volumetric injection flow rate to the pressure drop [55, 56]. For a homogeneous and isotropic reservoir, the steady-state CO_2 well injectivity index can be expressed as:

$$I = \frac{q}{\Delta p} = \frac{\rho_{CO_2,res}}{\rho_{CO_2,sc}} \frac{2\pi kh}{\left[\ln\left(\frac{r_e}{r_w}\right) + s\right]\mu_{CO_2}} \tag{3}$$

In Eq. (3), q is the volumetric injection flow rate, Δp is the pressure drop, $\rho_{CO_2,res}$ is the density of CO_2 under reservoir conditions, $\rho_{CO_2,sc}$ is the density of CO_2 under standard conditions, kh is the permeability-thickness product, r_e is the radius of the reservoir boundary, r_w is the well radius, s is the skin factor and $\mu_{CO_2}$is the viscosity of CO_2 under reservoir conditions. Well injectivity determines the number of wells required to inject a specific quantity of CO_2 into the reservoir. This makes injectivity an important factor for both technical and economic evaluation of CO_2 storage projects [56, 57].

4.3 Containment efficiency

Containment efficiency characterizes the assurance of containment of the injected CO_2. The ultimate objective of a CCUS project is to permanently isolate the sequestered CO_2 from the environment. Since formation water is denser than supercritical CO_2, the CO_2 plume tends to rise to the top of the reservoir, where it accumulates beneath the caprock. The containment efficiency of a geological trap is therefore strongly dependent on the seal potential or the ability of the caprock to confine the injected gas and prevent leakage into overlying formations and eventually back into the atmosphere [58]. The caprock must have the lateral extent and geomechanical strength to retain the full CO_2 column height.

The integrity of the caprock could be compromised by mechanical deformation induced by pressure from CO_2 injection or through geochemical CO_2-rock-brine interactions which may dissolve or precipitate minerals to increase the permeability of the caprock [59]. Wells have also been identified as probable leakage pathways. Therefore, robust wellbore integrity is important to prevent leakage through wells.

5. CO_2 well injectivity impairment mechanisms

CO_2 injectivity impairment is a technical and economic constraint on geological storage of CO_2 [12, 60, 61]. Under typical storage conditions, several factors could influence CO_2 injectivity because of the complex interplay of chemical and physical phenomena in the reservoir, especially in the injection area of the wellbore [46, 62, 63]. The injection area of the wellbore has the highest flactuations of temperature, pressure and flux, making it one of the most important sections of the formation for well injectivity impairment analysis. Lombard et al. [64] identified three main mechanisms responsible for CO_2 injectivity impairment: Geochemical, geomechanical and transport phenomena (**Figure 1**). Later Torsaeter et al. [65] did an extensive review on other CO_2 injectivity impairment mechanisms that they recommend should be given close attention in addition to salt precipitation effects. The geochemical mechanisms involve CO_2-brine-rock interactions, mineral dissolution, and precipitation. The transport effects include drying of the reservoir rock and fines mobilization. The geomechanical effects, which include borehole deformation has not been given enough research attention compared to the geochemical and transport mechanisms. These phenomena depend on the physical and chemical properties of injected CO_2 which are in turn driven by reservoir conditions and rock properties.

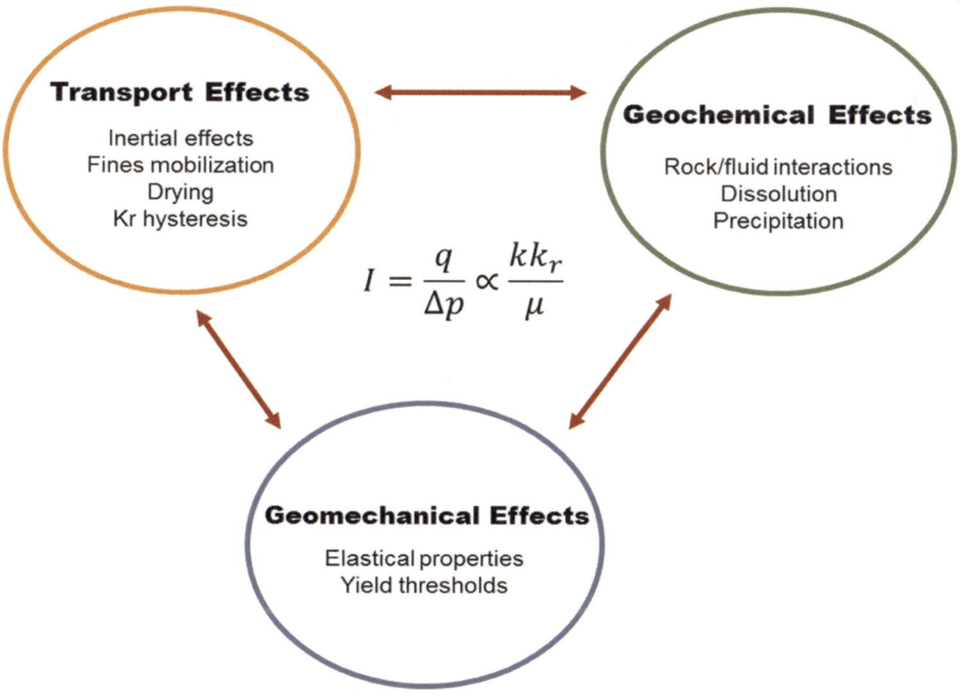

Figure 1.
CO_2 injectivity impairment mechanisms (after Lombard et al., [64]).

5.1 Geochemical effects

5.1.1 Effect of mineral dissolution

Mineral dissolution and salt precipitation are the two main geochemical CO_2 injectivity impairment mechanisms [64]. The injected CO_2 could dissolve in formation brine at the CO_2-brine interface, altering the concentration of aquifer fluid, thus leading to precipitation. Regardless of rock composition, the progressive dissolution of CO_2 in the brine (formation water) forms carbonic acid that leads to a reduction in pH to about 3–5 [66, 67]. The following reactions occur at the interface between both media

$$\begin{aligned} CO_{2(g)} + H_2O_{(l)} &\leftrightarrow H_2CO_{3(aq)} \\ H_2CO_{3(aq)} &\leftrightarrow H^+_{(aq)} + HCO^-_{3(aq)} \end{aligned} \tag{4}$$

Interactions between CO_2 and brine forms carbonic acid and then bicarbonates. In the presence of $CaCO_3$, $MgCO_3$ and $FeCO_3$, the following reactions would lead to formation of water-soluble bicarbonates.

$$\begin{aligned} H_2O + CO_2 + CaCO_3 &\rightleftarrows Ca(HCO_3)_2 \\ H_2O + CO_2 + MgCO_3 &\rightleftarrows Mg(HCO_3)_2 \\ H_2O + CO_2 + FeCO_3 &\rightleftarrows Fe(HCO_3)_2 \end{aligned} \tag{5}$$

Moreover, the bicarbonates could react with cations in the rock and formation water to form stable carbonates which later could aggregate into small particles or form a scale on the pore walls [68–70]. CO_2-brine-rock batch reaction under typical storage conditions have shown various amounts of dissolved minerals in solution [70–73].

Wang et al. [74] investigated mineral dissolution and precipitation in dolomite samples saturated with carbonated water at 93°C and 34.5 Mpa under static conditions. They found that mineral dissolution is predominant in highly permeable pathways in the reservoir rock. Zou et al. [75] conducted CO_2–brine–rock reactions under different conditions to investigate the changes of porosity and permeability during sCO_2 – fracturing in shale reservoir. They reported that CO_2–brine–rock reaction could occur rapidly (in less than 0.5 h), precipitating minerals such as calcite, dolomite, K-feldspar, and albite into solution. Zhang et al. [76] investigated CO_2-brine-rock reaction in a tight sandstone reservoir under conditions of 75°C and 32MPa. They found that the dissolution of supercritical CO_2 in brine created an acidic environment which induced the dissolution of minerals into the brine, leading to precipitation of iron minerals and kaolinite. Tang et al. [77] studied the impact of CO_2-brine-rock interaction on formation properties in gas zone and water zones of a reservoir under static and dynamic conditions. They found that CO_2-brine-rock interaction occurs in both gas zones and water zones. Aminu et al. [78] investigated the effect of CO_2 and the presence of impurities (NO_2, SO_2, H_2S) on reservoir permeability in static CO_2-brine-rock reactions. They found that the presence of impurities affacts the impact of mineral dissolution on rock permeability. Okhovat et al. [79] investigated the effect of CO_2-water-rock interaction on rock permeability and oil recovery during CO_2 injection into a carbonate rock. They found that the extent of damage induced was a function of the injection rate.

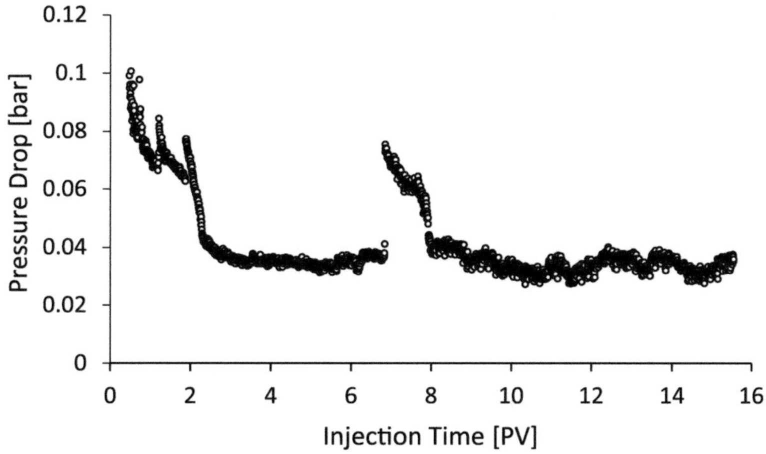

Figure 2.
Pressure drop profile recorded during injection of carbonated water into a Bentheimer core at flow rate of 0.25 mL/min at 80 bar and 60°C [80].

Under static conditions, mineral dissolution could increase rock permeability temporally as new pore spaces are etched and old pore channels could be widened [64]. However, mineral precipitates could aggregate into fine particles in the pore fluid which could form a scale on the pore walls and reduce the flow area.

Sokama-Neuyam et al. [80] conducted experiments to investigate the effect of mineral dissolution on CO_2 injectivity using clay-rich Bentheimer Sandstone cores. The cores were flooded with carbonated water at 80 bar and 60°C with about 25 pore volumes (PV) of carbonated water at 0.25 mL/min. Pressure drop profiles recorded during the flooding is shown in **Figure 2** and SEM-EDS analysis of effluent samples collected are shown in **Table 1**.

Figure 2 shows a period of immiscible displacement, where pressure drop decreased sharply from about 0.1 bar to about 0.04 bar after the core was flooded with

Element	wt.%
O	33.56
Fe	7.78
Ni	5.02
Na	17.53
Mg	0.74
Al	2.53
Si	0.35
Cl	29.79
Ca	2.52
Co	0.17
Total	100.00

Table 1.
EDS elemental analysis of effluent samples collected during carbonated water flooding into Bentheimer core.

about 3 PV of carbonated water. As the displaced FW is replaced with the less dense carbonated water, the pressure drop falls sharply until carbonated water breakthrough at the effluent end of the core. Pressure drop was stabilized from about 3 PV to about 7 PV. In this period, the core is fully saturated with carbonated water, leading to stable flow. Unstable flow sets in from about 7 PV to the end of the test. In this period, the pressure drop is seen to rise sharply and fall to a rather haphazard behavior towards the end of the test. The EDS results in **Table 1** shows the effluent sample is composed predominantly of Na (17.53%) and Cl (29.79) which are probably the components of the effluent brine. The results also reveal the presence of minerals such as Fe, Si, Al, Ni, and Co which were not present in the saturating brine and the injected carbonated water. Therefore, Fe, Si, Al, Ni, and Co were most likely dissolved from the Bentheimer core through the interaction of the carbonated water and the rock minerals. The EDS analysis show low amounts of Fe, Si, Al, Ni, and Co, because only few particles were likely to be washed out of the core.

5.1.2 Effect of salt precipitation

Salt precipitation is an existing injectivity challenge in natural gas wells. Kleinitz et al. [81] reported field observation of severe halite-scaling during natural gas production. Similar field experiences have been reported during injection, storage and production of natural gas [82–84]. In the context of field CO_2 injection, Baumann et al. [85] and Grude et al. [86] have reported evidence of salt precipitation effects in the Ketzin pilot reservoir and the Snøhvit field, respectively. More recently, Talman et al. [87] investigated drying and salt precipitation effects in a CO_2 injection well at the Aquistore site. Downhole images taken from the injection well, together with recovered samples revealed that scales of salts have formed on the inside of the injection well.

Miri and Hellevang [12] identified the processes leading to salt precipitation as: (1) immiscible two-phase CO_2-brine displacement; (2) vaporization of brine into the flowing CO_2 stream; (3) capillary back-flow of brine towards the inlet; (4) diffusion of dissolved salt in the porewater; (5) gravity override of CO_2; and (6) salt self-enhancing. While results from numerical modeling work reported by Roels et al. [88] suggested that precipitated salt could accumulate far from the wellbore, several research works [89–91] show that precipitated salt accumulates near the wellbore. Permeability impairment between 13 and 83% and porosity reduction between 2 and 15% have been reported from laboratory core-flood experiments [92–97]. These experimental findings have been found to be consistent with theoretical and numerical simulations [61, 98–100].

Pruess and Muller [89] suggested that pre-flush of the injection region with freshwater could reduce salt precipitation. However, Kleinitz et al., [81] have shown that freshwater injection could not mitigate salt precipitation if the flow area is completely plugged by solid salt. Fresh water also has a high tendency to react with rock minerals, leading to other injectivity impairment challenges such as clay swelling.

Sokama-Neuyam [101] grouped the mechanisms of salt precipitation into two successive processes: salt cake development at the injection inlet and drying effects. Salt cake forms on the surface of the core inlet during early stages of brine vaporization prior to drying. As drying commences, salt precipitates into pore spaces in the dry-out zone.

Sokama-Neuyam [101] investigated the development of salt cake on the surface of the injection inlet. They flooded a Bentheimer sandstone saturated with about 120 g/L

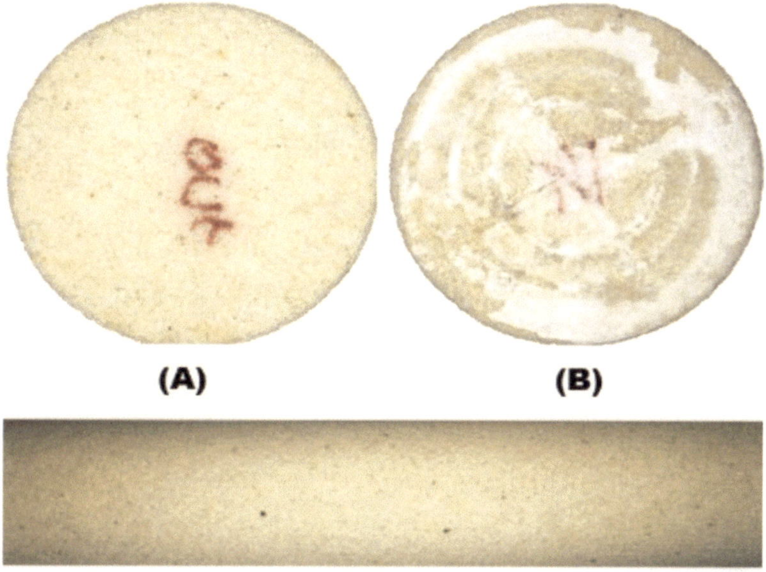

Figure 3.
Photographs of Bentheirmer core after CO_2 was injected at 1 mL/min into the core initially saturated with 120 g/L NaCl brine. (A) No salt cake observed at the core outlet. (B) Massive salt cake found at the injection inlet. (C) the entire core remains wet [101].

NaCl brine with about 100 PV of dry supercritical CO_2 at a rate of 1 mL/min. **Figure 3** shows photographs of the rock taken during inspection. In **Figure 3A**, we observe that no salt was formed at the core outlet. **Figure 3B** shows massive salt cake deposition at the core inlet and **Figure 3C** shows that the entire length of the core was still wet. They reported that salt cake formation at the injection inlet was caused by (1) High brine salinity and (2) Poor brine displacement at the injection inlet.

Furthermore, there are similarities between the increasing trend of injectivity impairment expressed by salt precipitation in Sokama-Neuyam [101] and those described by Yusof et al. [102, 103]. In their study, they examined the effects of brine salinity and brine type on CO_2 injectivity changes. They found that the injectivity reduction increased almost linearly between 6 and 27.3% as the brine salinity increases from zero to 100,000 ppm. The increasing growth of salt precipitation which reduced the porosity and effective flow area was identified as the main cause of the downtrend of CO_2 injectivity. It was also reported that the sandstone core saturated with monovalent salt such as NaCl and KCl was heavily impaired by salt precipitation as compared to the sandstone core filled by the divalent salt system ($CaCl_2$). However, these findings may be somewhat limited by constant brine salinity of 30,000 ppm.

Moreover, to investigate the effect of drying, Sokama-Neuyam [101] flooded a Berea sandstone core initially saturated with NaCl brine with about 300 PV of supercritical CO_2 at a rate of 1 mL/min until the core was completely dried. The permeability of the core after drying was measured and a relative injectivity indexβ which measures the injectivity of the core before and after impairment was calculated. They then repeated the test at injection flow rate of 5 mL/min and 10 mL/min, keeping all other parameters constant, to study the effect of injection flow rate. **Figure 4** shows

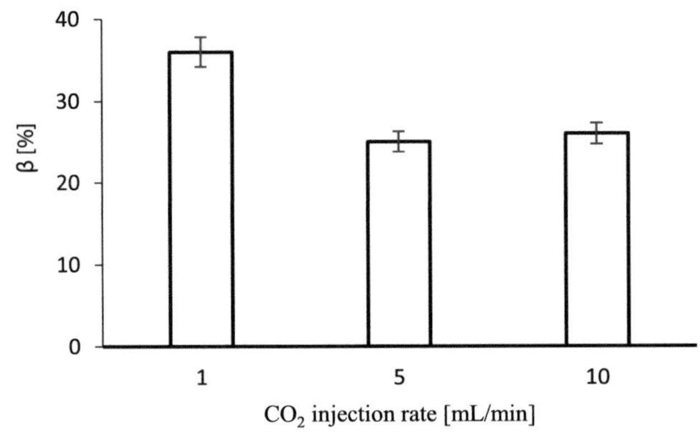

Figure 4.
Effect of drying and salt precipitation on CO_2 injectivity. Injectivity impairment, β increased with decreasing CO_2 injection rate.

results of injectivity impairment induced by drying and salt precipitation at varying CO_2 injection flow rates.

From **Figure 4**, it was observed that CO_2 injectivity was impaired by about 36% for drying rate of 1 mL/min. Injectivity impairment decreased from 36% to about 25% when drying rate was increased to 5 mL/min and remained practically unchanged when the drying rate was further increased to 10 mL/min. Several researchers [94, 96, 104, 105] have earlier reported CO_2 injectivity impairment within a range (13–83%) that is in agreement with the these figures. During drying and brine vaporization, when the concentration of brine exceed supersaturation, salt precipitates into the pores in the dry-out region as also observed by Zuluaga et al. [106]. The deposited salts reduce the CO_2 flow area, impairing permeability and injectivity.

As drying progresses, a saturation gradient is established which draws more brine into the dry-out region through capillary backflow. Capillary backflow of brine leads to more salt deposition in the dry-out region. The capillary backflow of brine increases with decreasing drying rate because at high CO_2 injection flow rates, viscous forces overcome capillary forces. Therefore, less salts are precipitated in the dry-out region at high injection flow rates, inducing low injectivity impairment as observed in **Figure 4**. Injectivity impairment did not change when drying rate was further increased from 5 mL/min to 10 mL/min because at these injection flow rates, the resident brine is quickly swept out of the core, leaving out only immobile brine for salt precipitation.

During injection of dry supercritical CO_2 into brine-saturated sandstone cores, the dry-out region close to the injection inlet, extends into the core as more CO_2 is injected. The effect of extension of the dry-out zone on CO_2 injectivity is vital for understanding the underlying mechanisms of brine vaporization and salt precipitation. Sokama-Neuyam et al. [107] conducted experimental and theoretical study to investigate the development of the dry-out zone and estimate the impact of extension of the dry-out region on CO_2 injectivity. **Figure 5** shows the impact of the advancing dry-out zone quantified by a dimensionless dry-out length, l_d on CO_2 injectivity impairment β. From **Figure 5**, CO_2 injectivity impairment peaked at the onset of drying. Injectivity impairment decreased to a minimum atl_d of about 0.45 and then rose slightly as the dry-out zone approached the core effluent end. At the

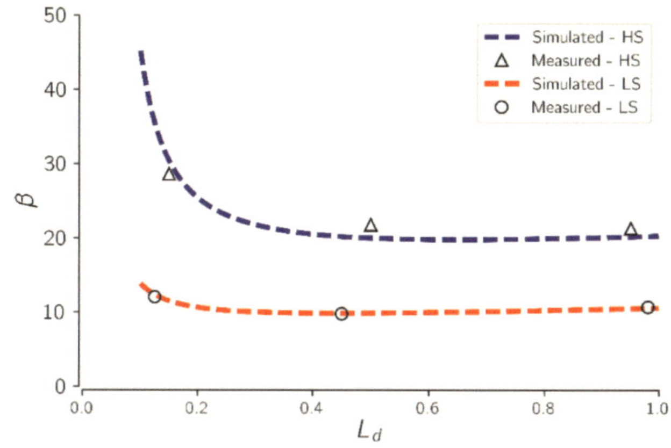

Figure 5.
The impact of the dry-out length (L_d) on CO_2 injectivity impairment (β) induced by salt precipitation. Magnitude of injectivity impairment increased when brine salinity was doubled from 75 g/L to 150 g/L but successive changes in injectivity impairment was not influenced by change in brine salinity [107].

start-up of drying, two mechanisms could impair CO_2 injectivity: salt precipitation and relative permeability effects. Brine vaporization rate is at its highest close to the inlet region because of the high capillary driven back-fluxes. As more brine is vaporized, more salt is precipitated into the pores which in turn increased the resistance to flow in this region. As the drying front advances into the core, brine vaporization and salt precipitation decreased as most of the brine were drawn into the inlet region by capillary backflow, leaving the remaining section of the core with less amount of brine available for salt precipitation. When the core is almost completely dried, brine vaporization and salt precipitation in around the effluent end of the core are almost negligible.

To meet the global CO_2 emission reduction target, large injection rates will be required. Thus, after salt precipitation, there is continuous injection of CO_2 into the reservoir. The effect of drag forces on the deposited salt was studied by Sokama-Neuyam et al. [108]. A Berea core sample was initially vacuum saturated with FW and vaporized to complete dryness to precipitate salt into the pores. The liquid CO_2 permeability of the core and pressure drop across two sections of the core were measured with a pressure-tapped core holder. The core was then flooded with about 150 PV of supercritical CO_2 at a constant injection rate of 2.5 mL/min. During this period of CO_2 injection, drag forces were expected to act on the precipitated salts. Permeability and pressure drop across the same sections of the core was measured after CO_2 flooding. **Figure 6** shows permeability change induced by the effect of drag forces on precipitated salts at varying injection rates.

The net drag force exerted by supercritical CO_2 on precipitated salts depends strongly on the volumetric injection flow rate (v), increasing with increase in flow rate. The precipitated salts are held to the pore walls mostly by gravitational and electrostatic forces offered by the complex pore structure and rock minerals. If drag forces overcome these attractive forces, the accumulated salts could be dislodged or redistributed in the pores, altering the permeability as a result. The magnitude of permeability change should therefore be proportional to the drag force which in tend depends on the injection flow rate.

CO_2 alternating Low Salinity Water Flooding (CO2-LSWAG) is a promising EOR technique [37, 38, 109, 110]. Sokama-Neuyam et al. [97] investigated CO_2 alternating

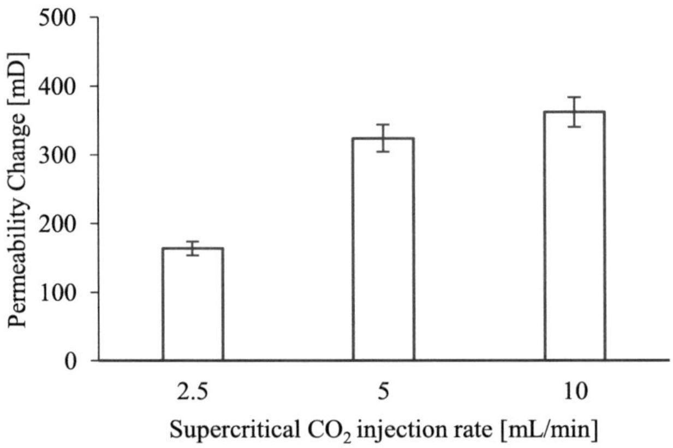

Figure 6.
The impact of CO_2 injection flow rate on the effect of drag on permeability after salt precipitation. Permeability change is the difference between the core permeability after salt precipitation and before drag test and the permeability after drag test [108].

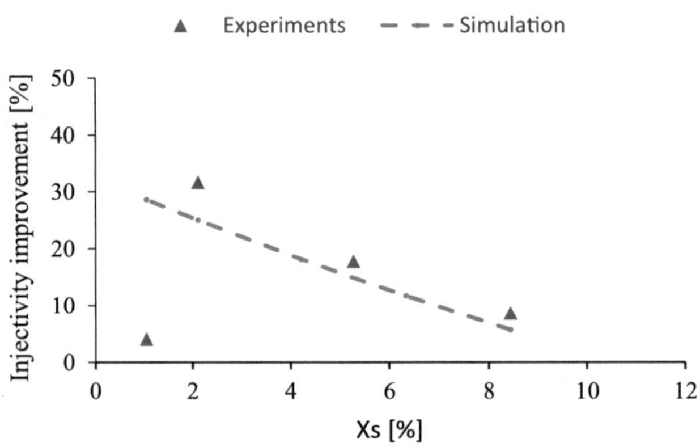

Figure 7.
Effect of diluent brine salinity on CO_2 injectivity change induced by alternate injection of supercritical CO_2 and LSW [111].

low salinity water injection as a potential technique to mitigate salt precipitation effects on CO_2 injectivity. After salt precipitation, a slug of diluent was injected to dissolve the precipitated salts, thus temporarily improving CO_2 injectivity. The diluent used was low salinity water (LSW) which was prepared by diluting FW to lower brine salinity. **Figure** 7 shows injectivity improvement obtained as a function of mass fraction of salt (X_s) in the diluent brine.

In general, CO_2 injectivity improved from 8.66 to 31.62% when the mass fraction of salt in the diluent,X_s was decreased stepwise from 8.44 to about 2.11. The solubility of precipitated minerals in the diluent LSW increased with decreasing brine salinity, because, as the brine is further diluted, more free water molecules become available to interact with the precipitated salts. At$X_s = 1.06$, injectivity dropped significantly and the experimental data deviates dramatically from the simulation results, signifying additional injectivity impairment other than salt precipitation. At this point, the diluent starts to interact chemically with the rock minerals. Interaction between rock minerals and the diluent could induce clay swelling and colloidal transport, which have the tendency to aggravate CO_2 injectivity impairment.

5.2 Transport effects

The mechanisms of fines migration and the impact of particle transport on the petrophysical properties of reservoir rocks have been previously researched. Khilar and Fogler [112] presented the mechanisms of colloidal and hydrodynamic induced release of fine particles in porous media. Muecke [113] investigated parameters controlling the movement of fine particles within the pore spaces. They identified the pH and salinity of formation brine, flow rate and temperature as some of the underlying parameters. Khilar and Fogler [114] asserted the existence of a critical salt concentration below which the pore fluid could weaken the Van der Waal's forces holding fine particles to the pore walls. Gruesbeck and Collins, [115] investigated the effect of hydrodynamic forces on the release and transport of fines. They identified a minimum interstitial velocity for fines entrainment. The effect of two-phase flow and rock wettability on fines entrainment has been experimentally investigated by Sarkar and Sharma [116]. They found that, the wettability of the core could affect the extent and rate of permeability impairment induced by migratory fines. Analytical models have been developed by Sharma and Yortsos [117] to investigate the mechanisms of size exclusion and quantify the effect of particle entrapment on rock permeability. Many other studies of fines migration in porous media under various conditions have been reported [118–120]. A thorough analysis of formation damage induced by migratory fines can be found in Civan [121].

We have already discussed that geochemical CO_2-brine-rock reaction could generate secondary minerals into the pore fluid [73, 122, 123]. In addition, CO_2-brine interactions could alter the pH of formation fluid which could induce the release of formation fines from the pore walls [115, 124]. While flowing with the injected fluid, the mineral particles could clog pore channels and impair injectivity. Whether entrapment or piping of fines will dominate the flow depends on characteristics of the generated fine particles, the porous medium and the permeating fluid in which the particles are suspended [125–127]. Pore structure, the size and concentration of the minerals and the hydrodynamic and colloidal conditions of the suspending medium could also affect their impact on CO_2 injectivity. Under radial flow conditions, plugging effects could be limited to the near well region where fluxes are highest.

The general mechanisms of fines mobilization in porous media have been well researched and understood. However, the unique properties of supercritical CO_2 including its gas-like viscosity and liquid-like density [128] coupled with the expected high CO_2 injection rates required to meet global emission reduction targets and the drying effect of supercritical CO_2 makes fines mobilization under CO_2 injection conditions a unique challenge that must be investigated separately. Adaptation and extension of previous general findings on fines migration is required to understand the mechanisms and impact of fines migration within the context of CO_2 injection.

Sokama-Neuyam [80] conducted core-flood experiments to measure the effect of dissolution on injectivity. A Berea sandstone core sample with known permeability was initially saturated with FW, and then flooded with about 25 PV of carbonated water at 80 bar and 60°C at constant injection rate of 0.25 mL/min to release and mobilize fine particles in the rock. The permeability of the core after carbonated water flooding was measured, and injectivity impairment index, β, was calculated. The experiment was then repeated at injection flow rates of 0.5 mL/min and 1.0 mL/min. **Figure 8** shows injectivity impairment decreased with increasing carbonated water injection flow rate.

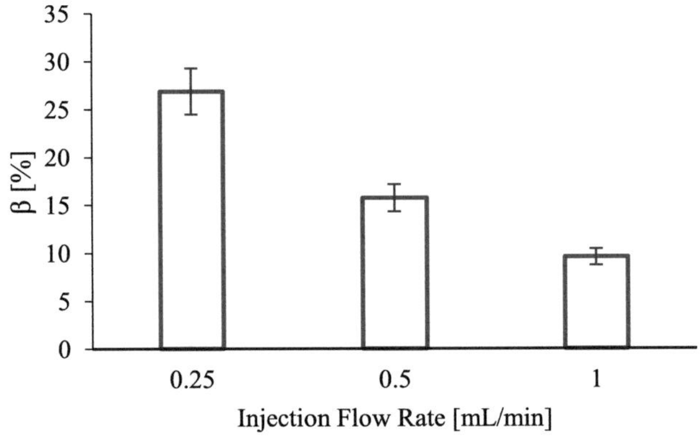

Figure 8.
Effect of injection flow rate on injectivity impairment induced by fines plugging. Injectivity impairment decreased with increasing carbonated water injection flow rate [80].

As injection flow rate is increased, the resident time of carbonated water in the rock is shortened. The number of fines generated will then decrease as the injection flow rate is increased. In addition, at high injection flow rate, hydrodynamic forces could lift smaller particles out of the core with the effluent fluid. The number of fines available to plug the rock, and therefore the chances of injectivity impairment, will reduce as carbonated water injection flow rate is increased. Up to 26% injectivity impairment was induced by mineral dissolution and fines mobilization during carbonated water injection into the Berea sandstone cores. Injectivity impairment decreased as injection flow rate was increased.

Sokama-Neuyam et al. [129] attempted to quantify and compare the individual effects of fines mobilization and salt precipitation on CO_2 injectivity. Mono-disperse colloid solutions were used to represent the pore fluid containing particles after mineral dissolution. A Berea sandstone core sample was initially saturated with mono-disperse colloid solution with average particle size of 0.08 μm and particle concentration of 0.3 wt.% and flooded with about 40 PV of supercritical CO_2 at 5 mL/min to complete dryness. The relative injectivity change, β, was calculated from the permeability of the core measured before and after it was exposed to mineral impairment. The experiment was repeated for particle concentrations of 0.5 and 1.0 wt.%. **Figure 9** shows injectivity impairment induced as a function of composition of the pore fluid.

In **Figure 9**, while salt precipitation reduced injectivity by about 26.8%, particle concentration of 0.3 wt.% impaired injectivity by 74.9% through fines mobilization. About 1.0 wt.% of particles in the pore fluid almost plugged the rock. When CO_2 invades the pores, the mono-disperse particles could plug the narrow pore channels through bridging, surface deposition, or multi-particle blocking. As particle concentration increases, the distance between suspended particles shortens, enhancing multi-particle blocking of the invaded pores. On the other hand, precipitated salts coat the pore walls to reduce the flow area. While salt precipitation reduces the flow area, fines entrapment could plug and isolate the flow path, making them inaccessible to fluid flow. The results suggest that, under linear flow conditions, fines mobilization could induce severe CO_2 injectivity impairment comparable to the impact of salt precipitation.

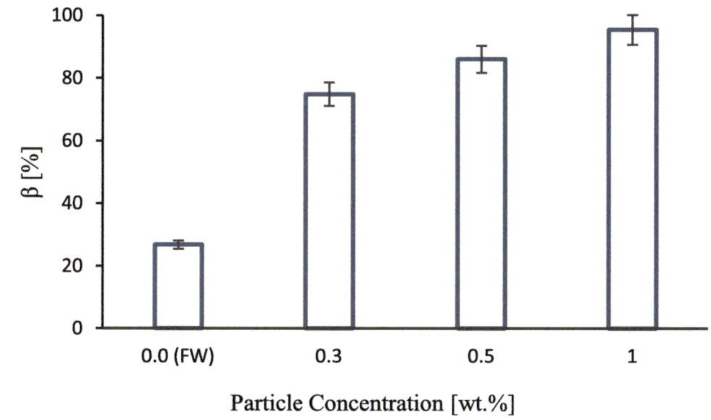

Figure 9.
The relative impact of fines mobilization and salt precipitation on CO2 injectivity. Fines migration had a more severe impact on injectivity compared to salt precipitation [129].

5.3 The coupled effect of salt precipitation and fines mobilization

Preliminary theoretical studies suggests that fines mobilization could compound CO_2 injectivity impairment induced by salt precipitation [80, 101]. After salt precipitation, the deposited salt reduces the pore spaces, increasing the jamming ratio of particles being transported in the flowing stream. A schematic diagram that summarizes the role of mineral dissolution, salt precipitation and fines migration mechanisms on CO_2 injectivity impairment is shown in **Figure 10**. Salt precipitation increases the susceptibility of the rock to fines entrapment. Sokama-Neuyam et al. [131], developed a dynamic core-scale model based on experimental observations to investigate the coupled effect of fines mobilization and salt precipitation on CO_2 injectivity. The effect of brine salinity, initial core permeability and the order of coupling were studied. **Figure 11** shows injectivity impairment induced by the combined effect of particle entrapment and salt precipitation compared to the effect of only fines mobilization.

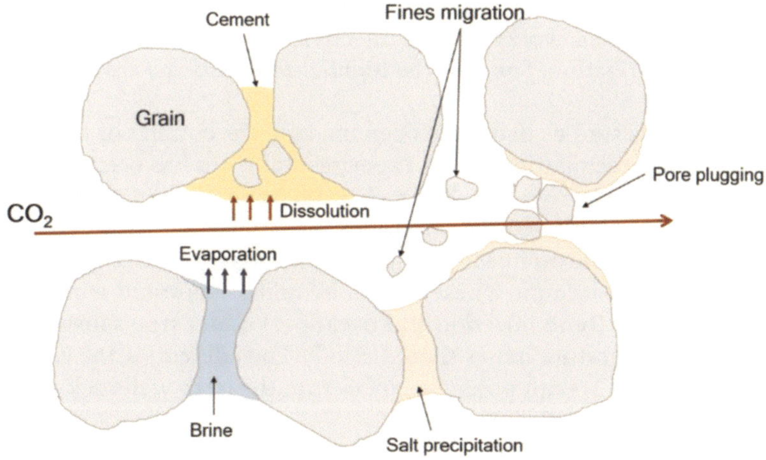

Figure 10.
Schematic diagram of mineral dissolution, salt precipitation and fines migration mechanisms during CO_2 injection into saline aquifer [130].

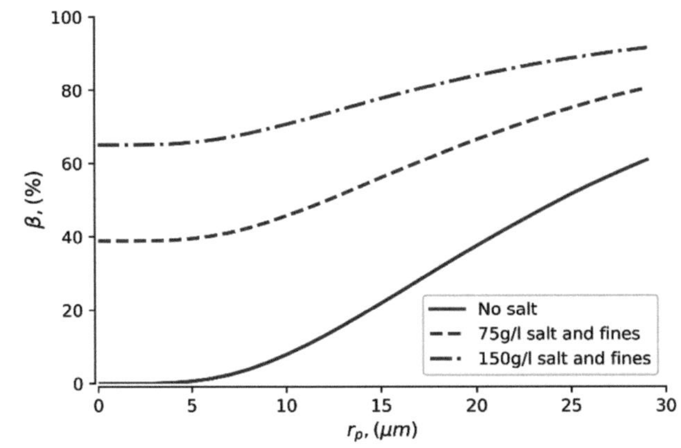

Figure 11.
Effect of fines mobilization compared to coupled effect of salt precipitation and particle entrapment [131].

Generally, injectivity impairment induced by particle entrapment increases with average particle size in the inlet fluid, r_p. For the base case (No salt), particles less than the average pore radius (6 μm) did not induce significant injectivity impairment. These particles generally piped through the pores as they have a jamming ratio less than 1.0. As r_p increases, more pores attain a jamming ratio greater than 1.0 and are plugged and isolated. As more pores are plugged and isolated, the flow area available to injected CO_2 decreases, impairing the injectivity. **Figure 11** also shows that coupling particles entrapment and salt precipitation increases CO_2 injectivity impairment significantly with the severity of impairment increasing with initial brine salinity. The deposited salt reduces the pore spaces and increases the jamming ratio. Particles that could pipe through the clean pores would be entrapped after salt precipitation. Thus, salt precipitation could compound the effect of fines mobilization on CO_2 injectivity.

6. Challenges and opportunities

Reviewing the previous works on CO_2 injectivity has thrown up many questions in need of further investigation. Some of the highlighted concerns are as follows.

- Most of previous studies that have been made in the context of CO_2 storage have focused on salt precipitation. Little experimental work has been performed to better understand near-well rock compaction, impact of temperature and operational parameters (drilling mud, residual hydrocarbons in pores). To give research-based advice on injectivity loss it is necessary to take geology and geomechanics into account. These cannot be reliably assessed without studying radial flow of CO_2 (both injection and backflow) under true subsurface stress conditions. Temperature issues should also be considered for the cases such as injection of cold CO_2 with potential to fracture the near-well rock and thus increase the permeability.

- Further work is required to enable direct pore-scale, real-time visualization of fluid-solid interactions with representative pore-geometry and realistic surface interactions between injectant, reservoir fluids and the formation rock.

- Current research approach to understand the reaction mechanism of CO_2-brine-rock for CO_2 sequestration are mainly limited to pure CO_2. However, due to high cost of gas purification, industrially sourced injection stream CO_2 contain impurities such as SO_2 which when dissolved in formation water will increase acidity beyond that of carbonic acid formed through CO_2 dissolution alone and may also have different reaction mechanism towards rock minerals present in subsurface CO_2 storage. Therefore, there is a need for a study to evaluate the effect of CO_2 and impurities in brine solution on chemical and physical rock properties.

- Majority of the dynamic CO_2 injection experiments are conducted within a limited time considering the limitation of available equipment. Therefore, there is an argument among the researchers on the sufficient CO_2 solubility into the brine and establishment of acidic environment that is critical in geochemistry study. Any future work to extend the CO_2 exposure time is highly recommended to identify the extended scale accumulation, mineral dissolution, and fines migration problems.

- Recent work on wettability during CO_2 injection has focused on the wettability changes of glass micromodel and shows that with increasing ionic strength, contact angle increases with increasing residence time. However, with understanding that fine particles released during CO_2 injection, may have not been exposed if the surface of the rock, its wettability may be slightly different or remain unaffected. Investigating the wettability of the released fines particles may provide insight on the behavior in which the fines particles migrate and accumulate.

- When it comes to mechanical stability, initial studies have shown that cyclic injection and shut-in may influence borehole deformation. This is probably exacerbated if some drawdown is experienced upon shut in. Shut-in cycle frequency effect on the amount of borehole deformation should be studied in more detail.

- Available field data and observation reports indicate the occurrence of injectivity loss. However, detail investigation analysis on the main issues that caused the problem is limited to certain fields. While there are more than 20 active fields worldwide, it should not be challenging to share their findings since the CO_2 mitigation would require collaboration.

7. Conclusion

In this chapter, a review of the current literature indicates that dissolution, precipitation, and fines mobilization are the main mechanisms that cause CO_2 injectivity impairments especially in deep saline reservoirs. Dissolution of carbonate minerals due to CO_2-brine-rock reaction is dominant and could increase the porosity and permeability of sandstone core samples. On the other hand, detachment, precipitation of salt and clay minerals and deposition of fines particles would decrease the permeability and even clog the flow paths despite net dissolution. The effect of these two seemingly opposing processes on CO_2 injectivity has been clearly demonstrated through numerous experimental studies supported by some field reports. However, the results are case dependent and lack generality in terms of quantifying the petrophysical damage.

There are many underlying parameters with positive and negative impacts on CO_2 injectivity. It has been highlighted that injection scheme (flow rate, time frame), mineral composition (clay content, sensitive minerals), particulate process in porous media (pore geometry, particle, and carrier fluid properties), and thermodynamic conditions (pressure, temperature, salinity, CO_2, and brine composition) have substantial effect on fines migration during CO_2 injection. However, there is abundant room for further progress in determining the impact of different fluid-rock mechanisms on CO_2 injectivity.

Author details

Yen A. Sokama-Neuyam[1]*, Muhammad A.M. Yusof[2] and Shadrack K. Owusu[1]

1 Department of Petroleum Engineering, Kwame Nkrumah University of Science and Technology, PMB, Kumasi, Ghana

2 Department of Petroleum Engineering, Universiti Teknologi PETRONAS, Seri Iskandar, Malaysia

*Address all correspondence to: asokama@knust.edu.gh

References

[1] Pachauri RK, Allen MR, Barros VR, Broome J, Cramer W, Christ R, et al. Climate change 2014: Synthesis report. Contribution of working groups I, II and III to the fifth assessment report of the intergovernmental panel on climate change. IPCC. 2014

[2] Shi Y, Wang C. Pore pressure generation in sedimentary basins: Overloading versus aquathermal. Journal of Geophysical Research—Solid Earth. 1986;**91**:2153-2162

[3] Baumert KA, Pershing J, Herzog T. Climate data: A sectoral perspective. Pew center on global. Climate Change. 2005

[4] Herzog T, Pershing J, Baumert KA. Navigating the Numbers. Washington D.C, USA: World Resources Institute; 2005

[5] Rogelj J, Den Elzen M, Höhne N, Fransen T, Fekete H, Winkler H, et al. Paris agreement climate proposals need a boost to keep warming well below 2 C. Nature. 2016;**534**:631-639

[6] Horowitz CA. Paris agreement. Int Leg Mater. 2016;**55**:740-755

[7] Savaresi A. The Paris agreement: A new beginning? Journal of Energy and Natural Resources Law. 2016;**34**:16-26

[8] Ritchie H, Roser M. CO_2 and greenhouse gas emissions. Our World Data. 2020

[9] Davison J, Freund P, Smith A. Putting carbon back in the ground. IEA Greenhouse Gas R & D Programme. 2001

[10] IEA. CO_2 Capture and Storage in Geological Formations. OECD/IEA. Paris, France: International Energy Agency; 2003

[11] Miri R. Effects of CO_2 -Brine-Rock Interactions on CO_2 Injectivity – Implications for CCS. Oslo: Uniersity of Oslo; 2015

[12] Miri R, Hellevang H. Salt precipitation during CO_2 storage—A review. International Journal of Greenh Gas Control. 2016;**51**:136-147. DOI: 10.1016/j.ijggc.2016.05.015

[13] Leung DYC, Caramanna G, Maroto-Valer MM. An overview of current status of carbon dioxide capture and storage technologies. Renewable and Sustainable Energy Reviews. 2014;**39**:426-443. DOI: 10.1016/j.rser.2014.07.093

[14] Huaman RNE, Jun TX. Energy related CO_2 emissions and the progress on CCS projects: A review. Renewable and Sustainable Energy Reviews. 2014; **31**:368-385

[15] Gibbins J, Chalmers H. Carbon capture and storage. Energy Policy. 2008;**36**:4317-4322

[16] IEA. Technology roadmap—Carbon capture and storage. Technol Roadmap. 2013:59. DOI: 10.1007/SpringerReference_7300

[17] Metz B, Davidson O, De Coninck HC, Loos M, Meyer L. IPCC Special Report on Carbon Dioxide Capture and Storage. Cambridge: Cambridge University Press; 2005

[18] Bentham M, Kirby M. CO_2 storage in saline aquifers. Oil & Gas Science and Technology. 2005;**60**:559-567

[19] Baines SJ, Worden RH. Geological Storage of Carbon Dioxide. London: Geological Society, Special Publications; 2009;**233**(1):1-6

[20] Gunter WD, Wong S, Cheel DB, Sjostrom G. Large CO_2 sinks: Their role in the mitigation of greenhouse gases from an international, national (Canadian) and provincial (Alberta) perspective. Applied Energy. 1998;**61**: 209-227

[21] Eccles JK, Pratson L, Newell RG, Jackson RB. Physical and economic potential of geological CO_2 storage in saline aquifers. Environmental Science & Technology. 2009;**43**:1962-1969

[22] Kumar A, Noh M, Pope GA, Sepehrnoori K, Bryant S, Lake LW. Reservoir Simulation of CO_2 Storage in Deep Saline Aquifers. SPE/DOE Symp. Improv. OnePetro: Oil Recover; 2004

[23] Singh N. Deep Saline Aquifiers for Sequestration of Carbon Dioxide. Oslo: International Geological Congress; 2008

[24] Potdar RS, Vishal V. Trapping mechanism of CO_2 storage in deep saline aquifers: Brief review. Geological Carbon Sequestration. 2016:47-58

[25] Reichle D, Houghton J, Kane B, Ekmann J. Carbon Sequestration Research and Development. Oak ridge National lab., TN (US); National Energy Technology lab, Pittsburgh, PA (US); National Energy Technology lab., Morgantown, WV (US); 1999

[26] Gozalpour F, Ren SR, Tohidi B. CO_2 Eor and Storage in Oil Reservoir Mendeley. 2005;**60**:537-546. DOI: 10.2516/ogst:2005036

[27] Núñez-López V, Moskal E. Potential of CO_2-EOR for near-term decarbonization. Frontiers in Climate. 2019;**1**:5

[28] Ettehadtavakkol A, Lake LW, Bryant SL. CO_2-EOR and storage design optimization. International Journal of Greenhouse Gas Control. 2014;**25**:79-92

[29] Rojas G, Ali SM. Scaled model studies of carbon dioxide/brine injection strategies for heavy oil recovery from thin formations. Journal of Canadian Petroleum Technology. 1986;**25**

[30] Tunio SQ, Tunio AH, Ghirano NA, El Adawy ZM. Comparison of different enhanced oil recovery techniques for better oil productivity. International Journal of Applied Science and Technology. 2011;**1**

[31] Jarrell PM. Practical aspects of CO_2 flooding. Richardson, Tex.: Henry L. Doherty Memorial Fund of AIME. Society of Petroleum Engineers. 2002

[32] Hill LB, Li X, Wei N. CO_2-EOR in China: A comparative review. International Journal of Greenh Gas Control. 2020;**103**:103173

[33] Talebian SH, Masoudi R, Tan IM, Zitha PLJ. Foam assisted CO_2-EOR: A review of concept, challenges, and future prospects. Journal of Petroleum Science and Engineering. 2014;**120**:202-215

[34] Esene C, Rezaei N, Aborig A, Zendehboudi S. Comprehensive review of carbonated water injection for enhanced oil recovery. Fuel. 2019;**237**: 1086-1107

[35] Bisweswar G, Al-Hamairi A, Jin S. Carbonated water injection: An efficient EOR approach. A review of fundamentals and prospects. Journal of Petroleum Exploration and Production Technologies. 2020;**10**:673-685

[36] Dang C, Nghiem L, Chen Z, Nguyen N, Nguyen Q. CO_2 Low Salinity Water Alternating Gas: a New Promising Approach for Enhanced Oil Recovery.

OnePetro: SPE Improv. Oil Recover. Symp; 2014

[37] Teklu TW, Alameri W, Graves RM, Kazemi H, AlSumaiti AM. Low-salinity water-alternating-CO_2 EOR. Journal of Petroleum Science and Engineering. 2016;**142**:101-118

[38] Chaturvedi KR, Ravilla D, Kaleem W, Jadhawar P, Sharma T. Impact of low salinity water injection on CO_2 storage and oil recovery for improved CO_2 utilization. Chemical Engineering Science. 2021;**229**:116127

[39] Stanton R, Flores R, Warwick PD, Gluskoter H, Stricker GD. Coal bed sequestration of carbon dioxide. In: 1st Natl. Conf. Carbon Sequestration. Washington, USA; 2001

[40] Perera MSA. Influences of CO_2 injection into deep coal seams: a review. Energy & Fuels. 2017;**31**:10324-10334

[41] Reeves SR. Enhanced CBM recovery, coalbed CO_2 sequestration assessed. Oil & Gas Journal. 2003;**101**:49-53

[42] IEA. Storing CO_2 in Unminable Coal Seams. IEAGHG n.d

[43] Jia B, Tsau J-S, Barati R. A review of the current progress of CO_2 injection EOR and carbon storage in shale oil reservoirs. Fuel. 2019;**236**:404-427

[44] Godec M, Koperna G, Petrusak R, Oudinot A. Enhanced gas recovery and CO_2 storage in gas shales: A summary review of its status and potential. Energy Procedia. 2014;**63**:5849-5857

[45] Rani S, Padmanabhan E, Prusty BK. Review of gas adsorption in shales for enhanced methane recovery and CO_2 storage. Journal of Petroleum Science and Engineering. 2019;**175**:634-643

[46] Cinar Y, Riaz A, Tchelepi HA. Experimental study of CO_2 injection into saline formations. In: SPE Annu. Tech. Conf. Exhib. Society of Petroleum Engineers; Washington D.C., USA: Society of Petroleum Engineers; 2007

[47] Yang F, Bai B, Tang D, Shari D-N, David W, Fang Y, et al. Characteristics of CO_2 sequestration in saline aquifers. Petroleum Science. 2010;**7**:83-92. DOI: 10.1007/s12182-010-0010-3

[48] Bachu S. Review of CO_2 storage efficiency in deep saline aquifers. International Journal of Greenh Gas Control. 2015;**40**:188-202. DOI: 10.1016/j.ijggc.2015.01.007

[49] Bachu S, Bonijoly D, Bradshaw J, Burruss R, Holloway S, Christensen NP, et al. CO_2 storage capacity estimation: Methodology and gaps. International Journal of Greenh Gas Control. 2007;**1**: 430-443. DOI: 10.1016/S1750-5836(07) 00086-2

[50] Bradshaw J, Bachu S, Bonijoly D, Burruss R, Holloway S, Christensen NP, et al. CO_2 storage capacity estimation: Issues and development of standards. International Journal of Greenh Gas Control. 2007;**1**:62-68. DOI: 10.1016/S1750-5836(07)00027-8

[51] De Silva PNK, Ranjith PG. A study of methodologies for CO_2storage capacity estimation of saline aquifers. Fuel. 2012; **93**:13-27. DOI: 10.1016/j.fuel.2011.07.004

[52] USDOE. Carbon sequestration ATLAS of the United States and Canada. In: US Dep Energy, off Foss Energy. National Energy Technol Lab; Washington D.C., USA: US Department of Energy; 2007

[53] Bachu S. Review of CO_2 storage efficiency in deep saline aquifers.

International Journal of Greenh gas Control. 2015;**40**. DOI: 10.1016/j.ijggc.2015.01.007

[54] Bachu S, Shaw JC, Pearson RM. Estimation of oil recovery and CO_2 storage capacity in CO_2 EOR incorporating the effect of underlying aquifers. SPE/DOE Fourteenth Symp Improv Oil Recover. 2004;**3**:1-13. DOI: 10.2523/89340-MS

[55] Dake LP. Fundamentals of Reservoir Engineering. Amsterdam, Netherlands: Elsevier; 1983

[56] Schembre-McCabe JM, Kamath J, Gurton RM. Mechanistic studies of CO_2 sequestration. International Petroleum Technology Conference. 2007. Conference paper number IPTC-11391-MS

[57] Birkholzer JT, Oldenburg CM, Zhou Q. CO_2 migration and pressure evolution in deep saline aquifers. International Journal of Greenh Gas Control. 2015;**40**:203-220

[58] Kaldi J, Daniel R, Tenthorey E, Michael K, Schacht U, Nicol A, et al. Containment of CO_2 in CCS: Role of Caprocks and faults. Energy Procedia. 2013;**37**:5403-5410. DOI: 10.1016/j.egypro.2013.06.458

[59] Daniel RF, Kaldi JG. Evaluating Seal Capacity of Cap Rocks and Intraformational Barriers for CO_2 Containment. Tulsa, Oklahoma, USA: American Association of Petroleum Geologists (AAPG); 2009

[60] Peysson Y, André L, Azaroual M. Well injectivity during CO_2 storage operations in deep saline aquifers-part 1: Experimental investigation of drying effects, salt precipitation and capillary forces. International Journal of Greenh Gas Control. 2014;**22**:291-300. DOI: 10.1016/j.ijggc.2013.10.031

[61] Pruess K. Formation dry-out from co2 injection into saline aquifers: 2. Analytical model for salt precipitation. Water Resources Research. 2009;**45**:1-6. DOI: 10.1029/2008WR007102

[62] Sundal A, Nystuen JP, Dypvik H, Miri R, Aagaard P. Effects of geological heterogeneity on CO_2 distribution and migration—A case study from the Johansen formation, Norway. Energy Procedia. 2013;**37**:5046-5054

[63] Cinar Y, South N, Riaz A, Tchelepi HA. Experimental study of CO_2 injection into saline formations. SPE Journal. 2009;**14**(4):588-594. DOI: 10.2523/110628-MS

[64] Lombard JM, Azaroual M, Pironon J, Broseta D, Egermann P, Munier G, et al. CO_2 injectivity in geological storages: An overview of program and results of the GeoCarbone-Injectivity project. Oil & Gas Science and Technology. 2010;**65**: 533-539. DOI: 10.2516/ogst/2010013

[65] Torsæter M, Cerasi P. International journal of greenhouse gas control geological and geomechanical factors impacting loss of near-well permeability during CO_2 injection. International Journal of Greenh Gas Control. 2018;**76**: 193-199. DOI: 10.1016/j.ijggc.2018.07.006

[66] Sigfusson B, Gislason SR, Matter JM, Stute M, Gunnlaugsson E, Gunnarsson I, et al. Solving the carbon-dioxide buoyancy challenge: The design and field testing of a dissolved CO_2 injection system. International Journal of Greenh Gas Control. 2015;**37**:213-219

[67] Md Yusof MA, Mohamed MA, Md Akhir NA, Ibrahim MA, Saaid IM, Idris AK, et al. Influence of brine–rock parameters on rock physical changes during CO_2 sequestration in Saline

Aquifer. Arabian Journal for Science and Engineering. 2021:1-15

[68] Patton JT, Phelan P, Holbrook S. CO_2 Formation Damage Study—First Annual Report. New Mexico: Las Cruces; 1981

[69] Sayegh SG, Krause FF, Girard M, DeBree C. Rock/fluid interactions of carbonated brines in a sandstone reservoir: Pembina Cardium, Alberta, Canada. SPE Form Eval. 1990;5:399-405. DOI: 10.2118/19392-PA

[70] Tobergte DR, Curtis S. Experimental perspectives of mineral dissolution and precipitation due to carbon dioxide-water-rock interactions. Journal of Chemical Information and Modeling. 2013;**53**:1689-1699. DOI: 10.1017/CBO9781107415324.004

[71] Dawson GKW, Pearce JK, Biddle D, Golding SD. Experimental mineral dissolution in Berea sandstone reacted with CO_2 or SO2-CO_2 in NaCl brine under CO_2 sequestration conditions. Chemical Geology. 2015;**399**:87-97. DOI: 10.1016/j.chemgeo.2014.10.005

[72] Kaszuba JP, Janecky DR, Snow MG. Experimental evaluation of mixed fluid reactions between supercritical carbon dioxide and NaCl brine: Relevance to the integrity of a geologic carbon repository. Chemical Geology. 2005;**217**:277-293. DOI: 10.1016/j.chemgeo.2004.12.014

[73] Ilgen AG, Cygan RT. Mineral dissolution and precipitation during CO_2 injection at the Frio-I brine pilot: Geochemical modeling and uncertainty analysis. International Journal of Greenh Gas Control. 2016;**44**:166-174. DOI: 10.1016/j.ijggc.2015.11.022

[74] Wang H, Alvarado V, Bagdonas DA, McLaughlin JF, Kaszuba JP, Grana D, et al. Effect of CO_2-brine-rock reactions on pore architecture and permeability in dolostone: Implications for CO_2 storage and EOR. International Journal of Greenh Gas Control. 2021;**107**:103283. DOI: 10.1016/j.ijggc.2021.103283

[75] Zou Y, Li S, Ma X, Zhang S, Li N, Chen M. Effects of CO_2–brine–rock interaction on porosity/permeability and mechanical properties during supercritical-CO2 fracturing in shale reservoirs. Journal of Natural Gas Science and Engineering. 2018;**49**: 157-168. DOI: 10.1016/j.jngse.2017.11.004

[76] Zhang X, Wei B, Shang J, Gao K, Pu W, Xu X, et al. Alterations of geochemical properties of a tight sandstone reservoir caused by supercritical CO_2-brine-rock interactions in CO_2-EOR and geosequestration. Journal of CO_2 Utilization. 2018;**28**: 408-418. DOI: 10.1016/j.jcou.2018.11.002

[77] Tang Y, Hu S, He Y, Wang Y, Wan X, Cui S, et al. Experiment on CO_2-brine-rock interaction during CO_2 injection and storage in gas reservoirs with aquifer. Chemical Engineering Journal. 2021;**413**:127567. DOI: 10.1016/j.cej.2020.127567

[78] Aminu MD, Nabavi SA, Manovic V. CO_2-brine-rock interactions: The effect of impurities on grain size distribution and reservoir permeability. International Journal of Greenh Gas Control. 2018;**78**: 168-176. DOI: 10.1016/j.ijggc.2018.08.008

[79] Okhovat MR, Hassani K, Rostami B, Khosravi M. Experimental studies of CO_2-brine-rock interaction effects on permeability alteration during CO_2-EOR. Journal of Petroleum Exploration and Production Technologies. 2020;**10**: 2293-2301. DOI: 10.1007/s13202-020-00883-8

[80] Sokama-Neuyam YA, Forsetløkken SL, Lien J, Ursin JR. The coupled effect of fines mobilization and salt precipitation on CO_2 Injectivity. Energies. 2017;**10**:1125. DOI: 10.3390/en10081125

[81] Kleinitz W, Dietzsch G, Köhler M. Halite scale formation in gas-producing wells. Chemical Engineering Research and Design. 2003;**81**:352-358

[82] Jasinski R, Sablerolle W, Amory M. ETAP: Scale Prediction and Control for the Heron Cluster. In: SPE Annu. Tech. Conf. Exhib. Society of Petroleum Engineers; Washington D.C. USA: Society of Petroleum Engineers. 1997

[83] Golghanddashti H, Saadat M, Abbasi S, Shahrabadi A. Experimental investigation of water vaporization and its induced formation damage associated with underground gas storage. Journal of Porous Media. 2013;**16**

[84] Place MC, Smith JT. An unusual case of salt plugging in a high-pressure sour gas well. In: SPE, editor. 59th SPE Annual Technical Conference and Exhibition. Houston, Texas: SPE; 1984. p. 13. DOI: 10.2118/13246-MS

[85] Baumann G, Henninges J, De Lucia M. Monitoring of saturation changes and salt precipitation during CO_2 injection using pulsed neutron-gamma logging at the Ketzin pilot site. International Journal of Greenh Gas Control. 2014;**28**:134-146

[86] Grude S, Landrø M, Dvorkin J. Pressure effects caused by CO_2 injection in the Tubåen Fm., the Snøhvit field. International Journal of Greenh Gas Control. 2014;**27**:178-187

[87] Talman S, Shokri AR, Chalaturnyk R, Nickel E. Salt precipitation at an active CO_2 injection site. Gas Inject into Geol Form Relat Top. 2020;**104**(1):22-28. DOI: 10.1002/9781119593324.ch11

[88] Roels SM, Ott H, Zitha PLJ. μ-CT analysis and numerical simulation of drying effects of CO_2 injection into brine-saturated porous media. International Journal of Greenh Gas Control. 2014;**27**:146-154. DOI: 10.1016/j.ijggc.2014.05.010

[89] Pruess K, Muller N. Formation dry-out from co2 injection into saline aquifers: 1. Effects of solids precipitation and their mitigation. Water Resources Research. 2009;**45**:1-11. DOI: 10.1029/2008WR007101

[90] Miri R, van Noort R, Aagaard P, Hellevang H. New insights on the physics of salt precipitation during injection of CO_2 into saline aquifers. International Jornal of Greenh Gas Control. 2015;**43**:10-21. DOI: 10.1016/j.ijggc.2015.10.004

[91] Sokama-Neuyam YAA, Ursin JR. The effect of mineral deposition on CO_2 well Injectivity. Europec. 2015;**2015**:1-4

[92] Bacci G, Korre A, Durucan S. Experimental investigation into salt precipitation during CO_2 injection in saline aquifers. Energy Procedia. 2011;**4**: 4450-4456. DOI: 10.1016/j.egypro.2011.02.399

[93] Kim M, Sell A, Sinton D. Aquifer-on-a-Chip: Understanding pore-scale salt precipitation dynamics during CO_2 sequestration. Lab on a Chip. 2013;**13**: 2508-2518. DOI: 10.1039/c3lc00031a

[94] Muller N, Qi R, Mackie E, Pruess K, Blunt MJ. CO_2 injection impairment due to halite precipitation. Energy Procedia. 2009;**1**:3507-3514. DOI: 10.1016/j.egypro.2009.02.143

[95] André L, Peysson Y, Azaroual M. Well injectivity during CO_2 storage operations in deep saline aquifers—Part 2 : Numerical simulations of drying , salt deposit mechanisms and role of capillary forces. Interantional Journal of Greenh Gas Control. 2014;**22**:301-312. DOI: 10.1016/j.ijggc.2013.10.030

[96] Tang Y, Yang R, Du Z, Zeng F. Experimental study of formation damage caused by complete water vaporization and salt precipitation in sandstone reservoirs. Transport in Porous Media. 2015;**107**:205-218

[97] Sokama-Neuyam YA, Ursin J. Experimental and theoretical investigations of CO_2 injectivity. AGH Drilling, Oil, Gas. 2016;**33**:245-257. DOI: 10.7494/drill.2016.33.2.245

[98] Giorgis T, Carpita M, Battistelli A. 2D modeling of salt precipitation during the injection of dry CO_2 in a depleted gas reservoir. Energy Conversion and Management. 2007;**48**:1816-1826. DOI: 10.1016/j.enconman.2007.01.012

[99] Hurter S, Berge J, Labregere D. Simulations for CO_2 injection projects with compositional simulator. Offshore Eur 4–7 Sept. 2007. DOI: 10.2118/108540-MS

[100] Zeidouni M, Pooladi-Darvish M, Keith D. Analytical solution to evaluate salt precipitation during CO_2 injection in saline aquifers. International Journal of Greenh Gas Control. 2009;**3**:600-611. DOI: 10.1016/j.ijggc.2009.04.004

[101] Sokama-Neuyam YA. Experimental and Theoretical Modelling of CO_2 Injectivity: Effect of Fines Migration and Salt Precipitation. Stavanger: University of Stavanger; 2017

[102] Md. Yusof MA, Ibrahim MA, Idress M, Idris AK, Saaid IM, Rosdi NM, et al. Effects of CO_2/rock/formation brine parameters on CO_2 injectivity for sequestration. SPE Journal. 2021;**26**: 1455-1468

[103] Md. Yusof MA, Ibrahim MA, Mohamed MA, Md Akhir NA, M Saaid I, Ziaudin Ahamed MN, et al. Predictive modelling of CO_2 injectivity impairment due to salt precipitation and fines migration during sequestration. In: International Petrolium Technology Conference. OnePetro; Washington D.C. USA: Society of Petroleum Engineers. 2021

[104] Bacci G, Durucan S, Korre A. Experimental and numerical study of the effects of halite scaling on Injectivity and seal performance during CO_2 injection in saline aquifers. Energy Procedia. 2013; **37**:3275-3282. DOI: 10.1016/j.egypro.2013.06.215

[105] Peysson Y. Permeability alteration induced by drying of brines in porous media. European Physical Journal Applied Physics. 2012;**60**:12. DOI: 10.1051/epjap/2012120088

[106] Zuluaga E, Muñoz NI, Obando GA. SPE 68335 An experimental study to evaluate water vaporisation and formation damage caused by dry gas flow through porous media. Media. 2001:SPE-68335-MS

[107] Sokama-neuyam YA, Ursin JR. Experimental and Theoretical Study of Salt Precipitation, Development of the Dry-out zone and CO_2 Injectivity. n.d:pp. 1-18

[108] Sokama-Neuyam YA, Ursin JR. CO_2 well Injectivity: Effect of viscous forces on precipitated minerals. International Petrolium Technology Conference. 2015; **13**. DOI: 10.2523/IPTC-18268-MS

[109] Wilson A. CO_2 low-salinity water alternating gas: A promising new

approach for EOR. Journal of Petroleum Technology. 2015;**67**. DOI: 10.2118/0115-0084-JPT

[110] Zolfaghari H, Zebarjadi A, Shahrokhi O, Ghazanfari MH. An experimental study of CO_2-low salinity water alternating gas injection in sandstone heavy oil reservoirs. Iran Journal of Oil Gas Science and Technology. 2013;**2**:37-47

[111] Sokama-neuyam YA, Ursin JR, Boakye P. Experimental investigation of the impact of salt precipitation on CO_2 injection in sandstone. 2019;**5**(1):1-9

[112] Khilar KC, Fogler HS. Migrations of Fines in Porous Media. Vol. 12. Berlin: Springer Science & Business Media; 1998

[113] Muecke TW. Formation fines and factors controlling their movement in porous media. Journal of Petroleum Technology. 1979;**31**:144-150

[114] Khilar KC, Fogler HS. Water sensitivity of sandstone. SPE. 1983;**23**(1):55-64

[115] Gruesbeck C, Collins RE. Entrainment and deposition of fine particles in porous. Media. 1982;**22**:10. DOI: 10.2118/8430-PA

[116] Sarkar A, Sharma M. Fines migration in two-phase flow. Journal of Petroleum Technology. 1990;**42**:646-652. DOI: 10.2118/17437-PA

[117] Sharma MM, Yortsos YC, Handy LL. Release and deposition of clays in sandstones. In SPE Oilfield and Geothermal Chemistry Symposium. OnePetro; 1985

[118] Han X, Zhong L, Liu Y, Fang T, Chen C. Experimental study and pore network modeling of formation damage induced by fines migration in unconsolidated sandstone reservoirs. Journal of Energy Resources Technology. 2020;**142**

[119] Prempeh K, Chequer L, Badalyan A, Bedrikovetsky P. Effects of kaolinite on fines migration and formation damage. In: SPE Int. Conf. Exhib. Form. Damage Control. OnePetro; 2020

[120] Giraldo LJ, Diez R, Acevedo S, Cortés FB, Franco CA. The effects of chemical composition of fines and nanoparticles on inhibition of formation damage caused by fines migration: Insights through a simplex-centroid mixture design of experiments. Journal of Petroleum Science and Engineering. 2021;**203**:108494

[121] Civan F. Reservoir Formation Damage. Gulf Professional Publishing; Amsterdam, Netherlands: Elsevier Inc.; 2015

[122] Wilkinson M, Haszeldine RS, Fallick AE, Odling N, Stoker SJ, Gatliff RW. CO_2-mineral reaction in a natural analogue for CO_2 storage—Implications for modeling. Journal of Sedimentary Research. 2009;**79**:486-494. DOI: 10.2110/jsr.2009.052

[123] Hangx SJT. Subsurface mineralisation: Rate of CO_2 mineralisation and geomechanical effects on host and seal formations. CATO Workpackage WP 4.1, Report. 2005:1-43

[124] Khilar KC, Fogler HS. Migrations of Fines in Porous Media. Netherlands: Springer; 2010

[125] Aji K. The Experimental and Theoretical Study of Fines Migration in Porous Media under Particle-Rock Repulsion and Attraction Kaiser Aji a Thesis Submitted for the Degree of Doctor of Philosophy (PhD) October 2014. 2014

[126] Sen TK, Khilar KC. Review on subsurface colloids and colloid-associated contaminant transport in saturated porous media. Advances in Colloid and Interface Science. 2006;**119**: 71-96. DOI: 10.1016/j.cis.2005.09.001

[127] Vaidya RN, Fogler HS. Formation damage due to colloidally induced fines migration. Colloids and Surfaces. 1990; **50**:215-229. DOI: 10.1016/0166-6622(90) 80265-6

[128] Nalawade SP, Picchioni F, Janssen LPBM. Supercritical carbon dioxide as a green solvent for processing polymer melts: Processing aspects and applications. Progress in Polymer Science. 2006;**31**:19-43. DOI: 10.1016/j. progpolymsci.2005.08.002

[129] Sokama-Neuyam YA, Ginting PUR, Timilsina B, Ursin JR. The impact of fines mobilization on CO_2 injectivity: An experimental study. International Journal of Greenh Gas Control. 2017;**65**. DOI: 10.1016/j.ijggc.2017.08.019

[130] Yusof MAM, Neuyam YAS, Ibrahim MA, Saaid IM, Idris AK, Mohamed MA. Experimental study of CO_2 injectivity impairment in sandstone due to salt precipitation and fines migration. Journal of Petroleum Exploration and Production Technologies. 2022. DOI: 10.1007/ s13202-022-01453-w

[131] Sokama-Neuyam YA, Ursin JR. The coupled effect of salt precipitation and fines mobilization on CO_2 injectivity in sandstone. Greenh Gases Science and Technology. 2018;**8**(6):1066-1078. DOI: 10.1002/ghg.1817

Chapter 6

Geomechanics of Geological Carbon Sequestration

Yongcun Feng and Shui Zhang

Abstract

Geological Carbon Sequestration (GCS) is an effective way to fight against global warming by capturing and injecting CO_2 into geological formations to ensure permanent storage as well as to prevent the environmental and health threats posed by carbon dioxide emissions. Security has been a key factor in the social acceptance of this technology, besides the issues related to economics. From a scientific point of view, the safety issues during CO_2 injection and long-term storage are highly related to geomechanics. This chapter provides a basic knowledge of the geomechanical issues involved in the GCS process to increase the understanding of safety issues and to improve the social acceptance of the technology among researchers and those interested in the technology.

Keywords: geological carbon sequestration, trapping mechanisms, stress change, caprock integrity, well integrity, induced seismicity

1. Introduction

To date, the application of GCS on a commercial scale is considered to be an effective solution for reducing the greenhouse effect [1]. GCS projects are carried out in highly permeable, porous formations at a certain depth, and CO_2 is typically injected in a supercritical state. It is necessary to ensure that the CO_2 injected into the subsurface does not or rarely leak within a very long time (at least 1000 years) for large-scale applications and public acceptance [2–6]. The types of CO_2 leaks are classified as physical and chemical and the essence of these is the geomechanical issues during CO_2 injection and storage, such as excessive stress changes, fault activation, damage to wellbore integrity, and caprock failure due to continuous injection and long-term storage [7–12]. This chapter introduces the geomechanical issues involved in the GCS process from the perspectives of the CO_2 trapping mechanism, in-situ stress changes, caprock performance, wellbore integrity, and induced seismicity.

2. Trapping mechanisms

In recent years, an increasing number of studies have focused on the short- and long-term effects of CO_2 injection into the subsurface. In most projects, CO_2 is injected in a supercritical state which can be stored in a gaseous, liquid, or supercritical state depending on the formation conditions. At the initial stage of injection, CO_2 will move toward the caprock layer due to the temperature-pressure conditions and density difference, and will eventually be blocked by the caprock layer. Then, the CO_2 can be captured as residual gas when groundwater intrusion occurs during the movement. In addition, CO_2 can be dissolved in groundwater and chemical reactions can occur by contact with rocks, which will contribute to CO_2 capture. Therefore, there are four trapping mechanisms for CO_2 in the storage process, that is, stratigraphic trapping, residual trapping, solubility trapping, and mineral trapping. These mechanisms are activated at different periods of the sequestration, as shown in **Figure 1**. Stratigraphic trapping is in charge of the initial CO_2 storage. Residual trapping and dissolved trapping play an important role in the transport of CO_2. Mineral trapping is formed when the CO_2 diffuses in the formation and contacts the rock. At this time, CO_2 is in the most stable state and the risk of leakage is minimized [14–18].

2.1 Stratigraphic trapping

The stratigraphic trapping mechanism is determined by geological structure [19]. A complex geological structure is formed during the deposition of the formation, and the locations with high and low permeability determine the fluid flow within the formation. The CO_2 injected into the formation will rise or move laterally until reaching a low permeable or impermeable caprock because the density of CO_2 is less than that of the fluid formation. CO_2 will be confined below the caprock in a supercritical, liquid, or gaseous state. Physical traps for storing CO_2 are formed by low-permeability formations or structures. The typical structural trap includes an anticline or a sealed fault, as shown in **Figure 2**.

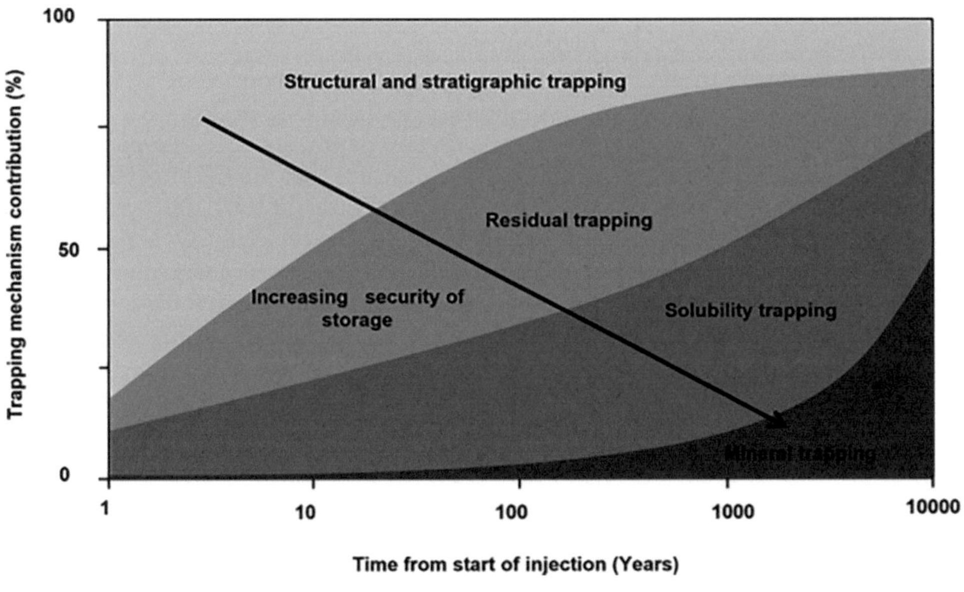

Figure 1.
Relative importance of trapping mechanisms with time [13].

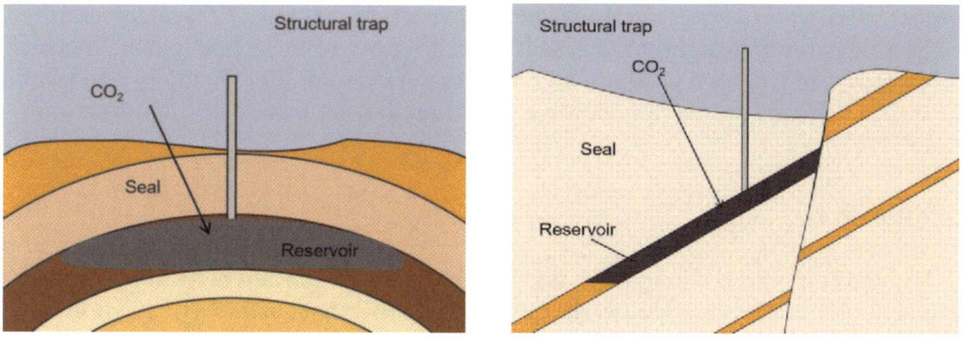

Figure 2.
Typical structural traps (reproduced from [20]).

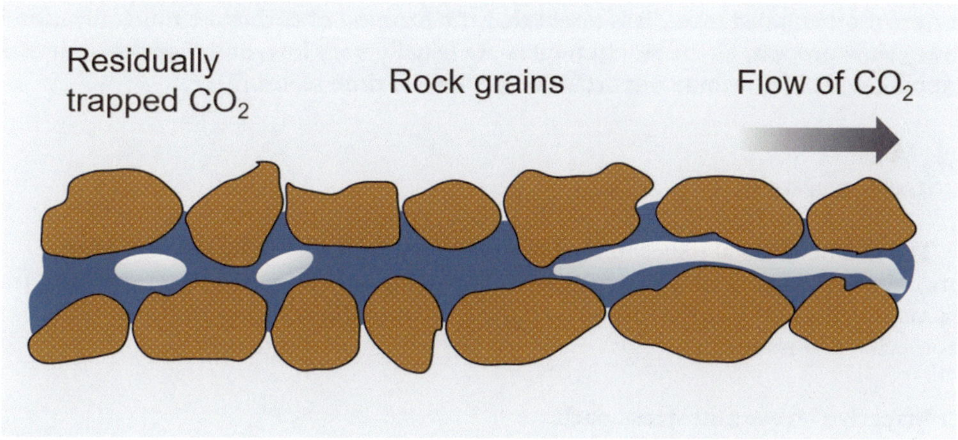

Figure 3.
Schematic diagram of residual trapping (reproduced from [20]).

2.2 Residual trapping

Residual trapping is a phenomenon in which CO_2 is trapped in the pores of rocks by capillary force. While the CO_2 is injected, it will enter the rock pore space and replace the original fluid. The difference in density between groundwater and CO_2 causes an upward. Then, the groundwater re-enters the rock pore space and the wetting phase (groundwater) will replace part of the weak wetting phase (CO_2). The replacement of CO_2 by groundwater leads to a significant reduction in the percentage of CO_2 in the rock pores, which are eventually trapped in the small pores, as shown in **Figure 3**. Thus, the isolated CO_2 is trapped as a stable phase by a trapping mechanism called residual trapping or capillary trapping [15, 21].

2.3 Solubility trapping

Solubility trapping is the dissolution of CO_2 in the formation fluid to achieve CO_2 storage. After injection, CO_2 is dissolved in the fluid formation until reaching saturation, due to the interaction of CO_2, groundwater, and hydrocarbons. The density difference between the fluid formation and CO_2 causes the CO_2 to migrate upward to contact more water formation that is not saturated. Meanwhile, CO_2 dissolved in the groundwater will slightly increase its density. Both of these phenomena increase the exchange of CO_2 and groundwater and accelerate solubility trapping. The solubility of CO_2 depends on the temperature, pressure, and saturation of the formation water [22].

2.4 Mineral trapping

Mineral trapping is a long-term trapping mechanism that involves contact and reaction with stratigraphic minerals and organic substances after CO_2 injection to form a stable mineral phase, resulting in long-term storage of CO_2. For example, the forming of carbonate minerals reduces the porosity and permeability of the rock and enhances the stability and integrity of the reservoir over time. The reaction rate of formation minerals with CO_2 depends on temperature, pressure, pH, and the concentration of other substances. It is noted that the forming of carbonate mineralization is a very slow process, as the reaction rates are usually very low, and therefore mineral trapping will only become important on geological time scales [23].

3. Stress response

The geomechanical issues in the GCS process are all driven by changes in the formation of pressure and ground stress. Therefore, it is important to first clarify the characteristics of the stress response in the formation for investigating the geomechanical issues.

3.1 Effective stress and stress path

3.1.1 Effective stress

The mechanical response of the rock is the result of the combination of pore pressure and in-situ stress. Terzaghi (1996) proposed the effective stress principle for describing the mechanical response of porous media. Effective stress is defined as the stress applied on the porous medium or the total stress minus the product of the pore pressure (fluid pressure) and the effective stress coefficient. In one-dimensional conditions it can be expressed as follows [24, 25]:

$$\sigma' = \sigma - \alpha P_p \tag{1}$$

where σ' is the effective stress; σ is the total stress; α is the effective stress coefficient; P_p is the pore pressure.

The three-dimensional condition is expressed as:

$$\sigma'_{ij} = \sigma_{ij} - \alpha P_p \delta_{ij} \tag{2}$$

where σ'_{ij} is the index notation of the effective stress tensor; σ_{ij} is the index notation of the total stress tensor; δ_{ij} is Kronecker's delta, when$i = j, \delta_{ij} = 1; i \neq j, \delta_{ij} = 0$.

The effective stress coefficient, also called the Biot coefficient, can be calculated by the following equation.

$$\alpha = 1 - \frac{K_{dry}}{K_m} \tag{3}$$

where K_{dry} is the bulk modulus of the dry porous rock; K_m is the bulk modulus of the matrix mineral in the rock. In Terzaghi's effective stress law α =1.

3.1.2 Stress path

The stress path, also known as the "stress history in the plane of maximum obliquity" is a common concept in geotechnics and rock mechanics. It refers to the trajectory of the stress path and stress history in the stress plane of stress space of a point in the core under the action of external forces, and is generally divided into effective stress path (ESP) and total stress path (TSP).

To understand the stress paths, consider a typical triaxial stress experiment in a core (**Figure 4a**). At any time, the stress state in the core can be represented by a Mohr circle (**Figure 4b**). It should be noted that during the triaxial experiments, the pore pressure can be neglected so that the total stress is equal to the effective stress. In triaxial compression tests, the maximum principal stress (σ_1) is applied along the axis of the cylindrical rock specimen, and the minimum principal stresses (σ_2 and σ_3) are applied on the lateral surface of the specimen. It is necessary to supplement the Mohr-Coulomb theory. Shear damage occurs at that point when the shear stress is equal to the shear strength of the material in any plane. The shear stress (shear strength) on the damaged plane depends on the normal stress on the shear plane and the properties of the rock and is a function of the normal stress on the shear plane [27–29].

The coordinates on the Mohr circle when considering the effective normal and shear stresses in the plane at an angle of 45° to the principal plane are calculated by [30–32]:

$$\text{The effective normal stress } p' = \frac{\sigma_1' + \sigma_3'}{2} \tag{4}$$

$$\text{The effective shear stress } q' = \frac{\sigma_1' - \sigma_3'}{2} \tag{5}$$

where σ_1' is the effective maximum principal stress; σ_3' is the effective minimum principal stress.

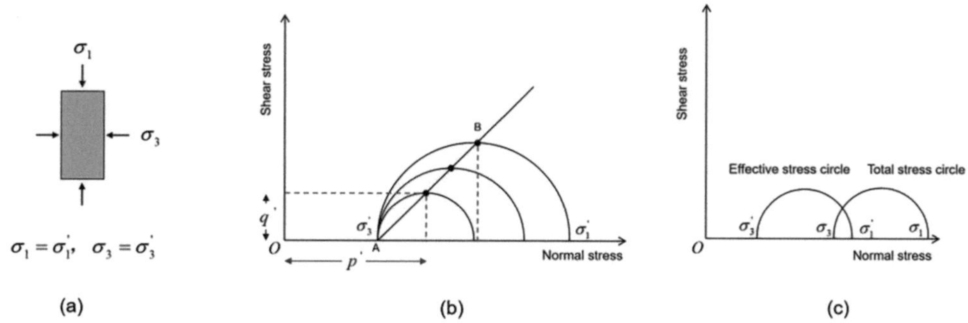

Figure 4.
(a) Triaxial stress experiment schematic; (b) laboratory stress path schematic; (c) schematic of total stress circle and effective stress circle (reproduced from [26]).

Connecting the points corresponding to coordinates p' and q' on each Mohr circle, as shown in line AB, is the stress path. In the formation conditions, the in-situ stress generally refers to the total stress. The pore pressure separates the effective stress circle from the total stress circle, as shown in **Figure 4c**, but they have the same diameter. In GCS engineering, shear failure, fault activation, and caprock failure can be determined by plotting the effective stress Mohr circle and effective stress path at a point in the formation [32].

3.1.3 Stress path coefficient

Effective stress is the key parameter for determining whether or not damage will occur in the rock. The injection of CO_2 will lead to an increase in the pore pressure. According to Eq. (1), the effective stress decreases, and the response to the Mohr circle is shifted to the left, as shown in **Figure 5**. Assuming constant total stress, the Mohr circle simply translates to the left until it intersects the failure envelope, resulting in shear damage. However, the increase in pore pressure leads to the expansion of the formation, which further leads to the change in total stress. The ratio of the variation of the total stress and the variation of the pore pressure is the stress path coefficient [33].

$$\gamma_v = \frac{\Delta\sigma_v}{\Delta p_p}, \gamma_h = \frac{\Delta\sigma_h}{\Delta p_p} \tag{6}$$

where γ_v is the vertical stress path coefficient; γ_h is the horizontal stress path coefficient; $\Delta\sigma_v$ is the vertical stress variation value; $\Delta\sigma_h$ is the horizontal stress variation value; Δp_p is the pore pressure variation value.

3.2 Effective stress variation law

During field construction, once fluid injection begins, the reservoir stress will change with the rapid propagation of fluid pressure in the injection zone, causing the

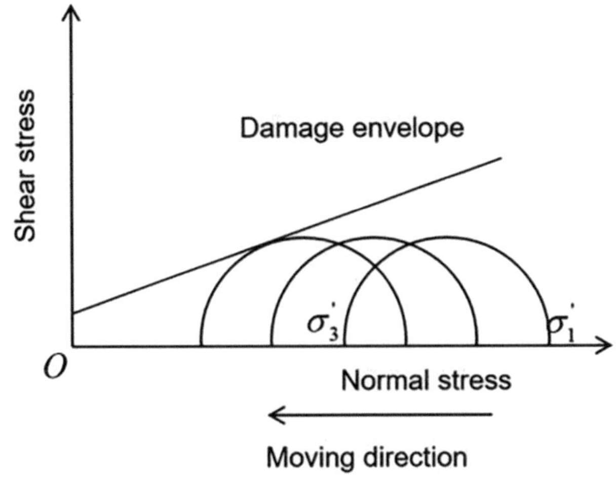

Figure 5.
Trend of Mohr circle with increasing pore pressure at constant total stress.

change in the reservoir stress field. In this case, the calculation of the effective stress requires consideration of the stress path coefficients. In a formation with a wide lateral distribution, the horizontal stress path coefficient can be calculated as [33, 34]:

$$\gamma_h = \frac{\Delta\sigma_h}{\Delta p_p} = \alpha\frac{1-2\mu}{1-\mu} \tag{7}$$

where μ is the Poisson's ratio.

The horizontal stress path coefficient is less than 1 because is less than 1. However, the vertical total stress can be considered to remain constant since the vertical formation expansion is not constrained, that is, the vertical stress path coefficient is equal to 0. It means that the horizontal total stress will change, while the vertical total stress remains constant. At this time the maximum and minimum effective stress change differently, Mohr's circle will not only move but also the diameter will change, as shown in **Figure 6**.

4. Caprock integrity

The effective and shear stress in the formation continues to change during the injection of CO_2. The tensile or shear damage will occur when the effective or shear stress reaches a certain critical point, forming fractures and providing leakage channels for CO_2.

4.1 Failure type

Two types of failures can occur during the continuous injection process, that is, tensile and shear failure. The effective stress continues to decrease to 0 as the pore pressure increases during the continuous injection of CO_2. Then it grows in the opposite direction and changes from compressive stress to tensile stress. Tensile failure will occur when the tensile stress exceeds the tensile strength of the rock. The principle is the same as that of hydraulic fracturing. Therefore, it is necessary to calculate the fracture pressure of the formation before injection, and then inject CO_2 at a pressure lower than the fracture pressure. Typical fracture pressure calculation methods can be found in hydraulic fracturing-related studies. The shear failure will

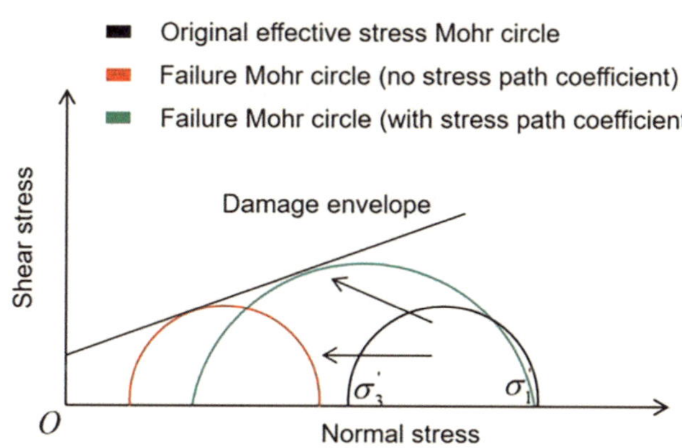

Figure 6.
Mohr circle variation considering stress path coefficients.

occur when the shear stress at a point reaches its shear strength. The Mohr-Coulomb criterion is used to determine whether a shear failure has occurred in the rock.

The maturity of hydraulic fracturing technology has made the determination of tensile failure easier. And the existence of stress path coefficients leads to a more complicated determination of shear damage. More importantly, the shear effect of injecting CO_2 varies in different fault regimes [35–39].

4.2 Normal fault regime

In a normal fault regime formation, as shown in **Figure 7a**, the maximum principal stress is the vertical (overburden) stress and the minimum principal stress is the minimum horizontal principal stress. In this stress regime, the initial Mohr circle of the formation is shown in **Figure 7b** and the variation of the Mohr circle with the injection of CO_2 is shown in **Figure 7c**. The variation of the maximum effective stress is greater as compared to that of the minimum effective stress when the pore pressure increases. The phenomenon that can be observed is that the radius of the Mohr circle will decrease and move slightly to the left, which implies the reduction of the shear stress. Therefore, at the early stage of CO_2 injection, the Mohr circle will move away from the damage envelope and the stability of the caprock and fault will increase to some extent. The shear stress decreases continuously with the injection of CO_2 until the maximum and minimum effective stresses are equal, and then the Mohr circle reverses [40, 41].

4.3 Strike-slip fault regime

In the strike-slip fault regime as shown in **Figure 8a**, both the maximum and minimum principal stresses are in the horizontal direction. In this regime, the initial

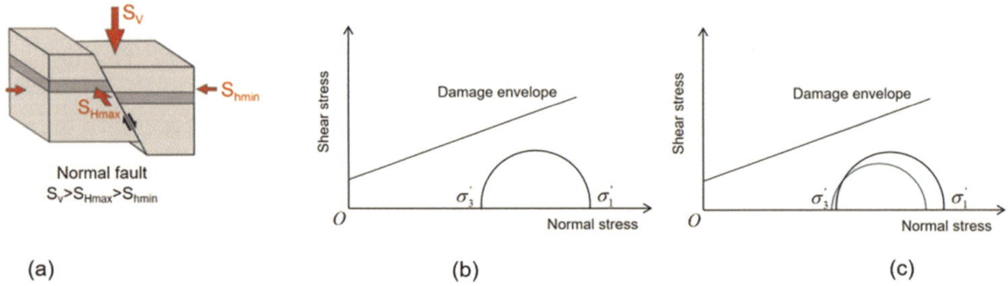

Figure 7.
(a) Normal fault regime; (b) initial Mohr circle; (c) schematic of Mohr circle variation.

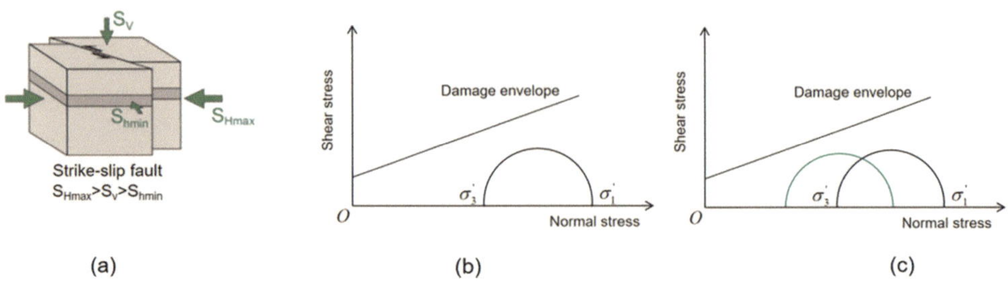

Figure 8.
(a) Strike-slip fault regime; (b) initial Mohr circle; (c) schematic of Mohr circle variation.

Mohr circle of the formation is shown in **Figure 8b**. The maximum and minimum effective stresses have the same variation and the diameter of the Mohr circle is constant when CO_2 is injected into the strike-slip regime formation because the horizontal stress path coefficients are the same. However, the Mohr circle rapidly moves to the left and closer to the damage envelope due to the reduction of the effective stress, causing the formation to become unstable, as shown in **Figure 8c** [42].

4.4 Thrust fault regime

In a thrust fault regime, as shown in **Figure 9a**, the maximum principal stress is the horizontal stress, while the minimum principal stress is the vertical stress. In this regime, the initial Mohr circle of the formation is shown in **Figure 9b**. The maximum effective stress changes less than the minimum effective stress when CO_2 is injected. Therefore, the radius of the Mohr circle increases and rapidly moves to the left during the injection process, and the shear stress increases (**Figure 9c**). The distance between the Mohr circle and the damage envelope will decrease rapidly, and then the caprock stability will decrease. Therefore, according to the trend of the Mohr circle, the reservoir and caprock will be more stable in the normal fault regime than in the thrust regime [40, 43].

In summary, the changes in Mohr circles are very different when CO_2 is injected into formations with different stress regimes. Generally, the formation is most stable in the normal fault regime, followed by the strike-slip regime, and the least stable is the thrust regime.

5. Well integrity

Well integrity is generally defined as the ability of a well to produce or inject a fluid while preventing harmful fluid leaks to reduce the risk of uncontrolled leakage of fluids formation throughout the life of the well. Well integrity is key to the success of geologic carbon sequestration and CO_2 enhanced recovery operations. Modern wells are designed with multiple barriers to create a controlled injection or production pathway and to isolate the fluid in the formation along with its depth. Wells will be impacted by physical, chemical, and mechanical stresses, which can reduce the effectiveness of the barrier and ultimately lead to loss of integrity. In CO_2 injection areas, these effects may be amplified because of high temperature and pressure variations in the injection well, higher injection pressure, and reactions between the CO_2-brine

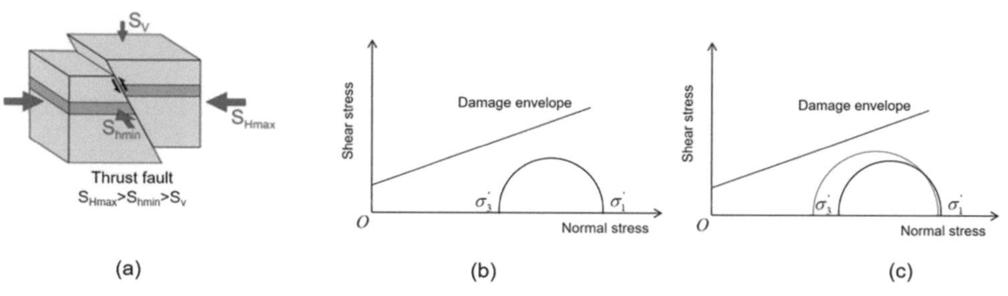

Figure 9.
(a) Thrust fault regime; (b) initial Mohr circle; (c) schematic of Mohr circle variation.

mixture with the well material. A well with compromised integrity may not be able to prevent upward transport of injected CO_2 and other formation fluids (e.g. brine, hydrocarbons). The leaked liquid will become greenhouse gas if it is released into the atmosphere and will contaminate drinkable groundwater resources if it leaks into subsurface water formations [44, 45].

5.1 Leak path

The function of cement sheath in oil and gas wells is to seal formation fluids, support and suspend casing, protect the wellbore, and provide a certain alkaline environment to avoid casing corrosion and ensure wellbore integrity. However, after carbon dioxide is injected into the subsurface, it will corrode the cement sheath and casing under suitable humidity and pressure conditions cause problems, such as a decrease in the strength of the cement sheath and an increase in its permeability. Furthermore, the corrosion effect, cement hardening, and alternate injection processes will impact casing and cement sheath strength and stress distribution in the near-well region, exacerbating wellbore integrity failure. **Figure 10** shows the CO_2 leak path due to wellbore integrity failure [45, 47].

(1) Casing corrosion creating fractures (#1)

(2) Incomplete and inadequate cement pouring existing leakage paths (#2, #9)

(3) Flow upward through the annulus or naked well (#3, #4)

(4) Leakage along the thread of the casing connection (#5)

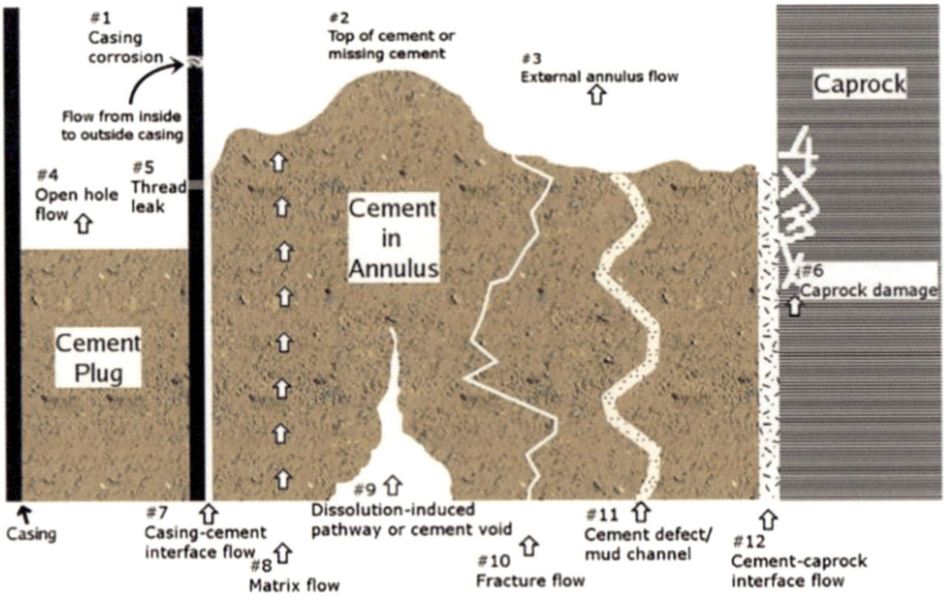

Figure 10.
Schematic diagram of leak pathways related to wellbore integrity (reproduced from [46]).

(5) Leakage along formations damaged during drilling (#6)

(6) Casing-cement and cement-formation interface debonding resulting in microfractures (#7, #12)

(7) Poor cement consolidation quality with permeability (#8)

(8) Fractures or gas channels in cement (#10, #11)

5.2 Chemical well integrity failure

Silicate cement is usually used as a sealing material when completing a well. The reaction occurs when the cement comes into contact with CO_2, resulting in the degradation of the cement. The continuous reaction of cement with CO_2 increases the porosity of the cement matrix, which also allows chloride ions to pass through the matrix causing corrosion of the casing [48, 49].

However, the consequences of the reaction between cement and CO_2 are not always harmful. The slight carbonation of CO_2 can reduce the permeability and porosity of the cement without causing failure of well integrity if the cement consolidation quality is excellent. However, a high degree of carbonation will certainly lead to failure or fracturing of the cement structure, failing well integrity. The corrosive effect of CO_2 will be more pronounced if the cement quality is poor with existing defects such as tiny pores or fractures [50, 51].

There are numerous mechanisms of chemical reactions impacting well integrity, and it is necessary to clarify the CO_2 and formation properties, and then the various chemical changes during the interaction of cement materials, wellbore materials, and carbonic acid can be further investigated.

5.3 Mechanical well integrity failure

Geomechanics is a key factor impacting well integrity with effects throughout the life cycle of a GCS project. During the drilling, completion, and application of the well the stresses in the wellbore, the cement sheath, and the near-well area keep changing. The reasons include pore pressure changes due to injection, leakage, and diffusion, thermal stress changes due to temperature differences, and in-situ stress changes due to tectonic shifts or seismic activity. Among them, the stress change in the cement sheath is the most significant. After the cement is poured, it will go through two stages of hardening (liquid cement transforms into solid cement) and shrinkage (solid cement volume shrinks and continues to harden). The change in cement rheology is more obvious during the hardening process. However, in the post-hardening phase, the shrinkage of the cement not only has a large effect on the stresses (because it is a solid shrinkage), but also the shrinkage may lead to plastic deformation or cause debonding at the interface between the cement and the formation or casing. In addition, the casing position may offset from the center of the wellbore, resulting in uneven stresses on the casing and cement sheath, increasing the risk of casing deformation and interface debonding. Thermal stresses are particularly important when CO_2 is injected into the subsurface, especially in the case of cyclic loading, causing coupling between thermal stresses and other stress fields, promoting fracture growth within the cement or debonding at the cement interface [46, 52, 53].

In conclusion, well integrity issues may be encountered during the entire GCS life cycle due to the influence of multiple factors such as coupled stress-thermal-chemical.

Therefore, it is necessary to take into account the combined effect of these factors in GCS engineering for well integrity problems, which are generally studied using numerical simulation.

6. Induced seismicity

From a safety and public perception point of view, a more unacceptable issue for CO_2 injection into the underground formation is the microseismic activity caused by fault activity or surface uplift during the injection process. Sometimes the intensity of induced seismicity will be perceived by humans. Slight shear sliding of fractures is induced when the induced seismic magnitude is very low. It will be more beneficial for CO_2 injection if the fractures are confined within the reservoir, as fracture sliding enhances permeability. However, microseismicity caused by GCS is difficult to control. Seismic events which can be sensed may have serious consequences such as massive CO_2 leakage, damage to injection wells, vertical surface displacement damaging buildings or infrastructure, etc. For example, several of the largest seismic events in the United States in 2011 and 2012 may have been caused by nearby disposal wells. The largest of these was a 5.6 magnitude seismic that occurred in Oklahoma, destroyed 14 buildings and injured two people. For the public, a perceptible seismic event would cause serious panic. Therefore, it is important to minimize or avoid seismic activity to ensure that geological energy projects such as GCS are carried out [39, 54–56].

6.1 Key factors affecting induced seismicity

As shown in **Figure 11** [54], the mechanism for inducing seismicity appears to be well known, that is, weakening pre-existing faults by increasing fluid pressure. The formation will release stored elastic strain energy when a fault slips, triggering seismicity. The fault will remain locked as long as the applied shear stress is less than the

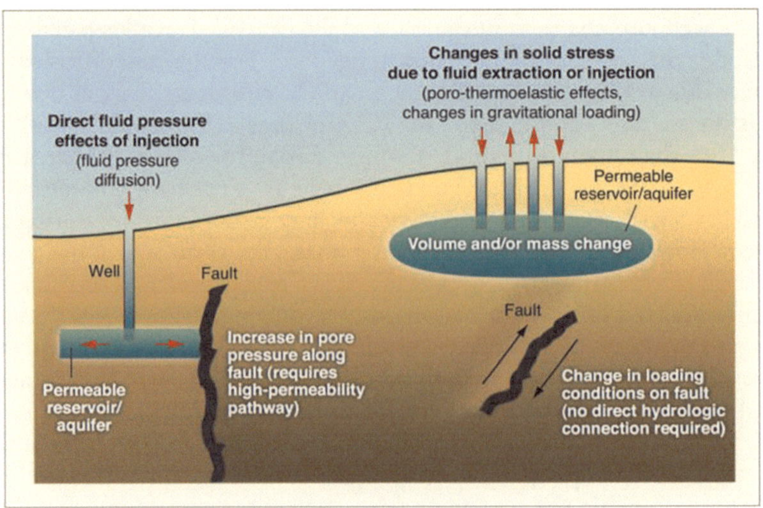

Figure 11.
Schematic diagram of mechanisms for inducing seismicity [54].

strength of the contact. The failure condition can be determined according to the effective stress principle and the Mohr-Coulomb criterion.

Theoretically, the increase of fluid pressure should be the direct reason for induced seismicity and it can be avoided by changing the injection rate and controlling the formation pressure to avoid induced seismicity. Based on the mechanism of induced seismicity, the possible reasons for seismic events caused by GCS are clarified [57–59].

1. The pore pressure increases and the stress state changes when the fluid is injected.

2. There is usually a temperature difference between injected CO_2 and the formation, which cools the formation near the injection well, causing thermal stress. The magnitude of thermal stress is proportional to the stiffness of the rock, so the thermal stress is more obvious when CO_2 is injected into a hard formation.

3. The parameter differences between the reservoir and the caprock result in different responses to pressure accumulation and thermal stress.

4. Each microseismic event will result in shear slip as well as stress redistribution. However, not all shear slips and the redistribution of stresses will cause microseismic events.

5. Geochemistry, rock strength, fault strength, and heterogeneity of stress field are all potential factors leading to local stress changes and triggering microseismicity.

6.2 Seismic moment release

At present, the principle and calculation formula of subsurface injection-induced seismicity proposed by McGarr [60] are widely used. The theory is based on the following assumptions to calculate the upper bound seismic moment due to fluid injection into geological formations.

1. There are seismogenic faults in the vicinity of the injection formation that are prone to slip in the contemporaneous state of stress.

2. The stress on the fault is within the seismic stress drop of failure due to earlier seismic activity.

3. The seismic rock mass is fully saturated before injection.

4. The induced seismicity is confined to areas weakened by fluid injection.

If a volume ΔV of liquid is injected into a fully saturated formation, the average increase in pore pressure P can be calculated as follows.

$$\Delta P = \frac{3\lambda + 2G}{3} \frac{\Delta V}{V} \tag{8}$$

where V is the volume of the formation weakened by the injection; λ and G are Lame's elastic parameters, G being the modulus of rigidity.

And the upper bound to the cumulative seismic moment is given by:

$$\sum M_0 = \frac{2\eta(3\lambda + 2G)}{3} \Delta V \tag{9}$$

where η is the coefficient of friction.

The analytical solutions that can accurately calculate the seismic magnitude are difficult to obtain because of the complexity of the induced seismicity. Therefore, more numerical simulations are used for seismic prediction, including the calculation of activation potential using shear slip criterion, based on continuous medium mechanics approach, through fault hydrodynamic approach, and discrete method modeling approach. A seismic event of magnitude 3–4 with a radius of the damage zone between a few hundred meters and one kilometer of the shallow formation can be perceived based on field experience and extensive numerical simulation studies. It is further demonstrated that high-level seismicity is induced only when the fault has continuous permeability and the pressure is distributed over a sufficiently large fault with simultaneous brittle fracture. In fact, even clearly perceivable seismic events may not open new flow channels over the entire thickness of the caprock, that is, seismically induced flow channels are unlikely to cross a formation with multiple caprocks.

7. Conclusions

GCS is an important way to reduce carbon emissions. There are multiple trapping mechanisms after CO_2 injection into the subsurface, including stratigraphic trapping, residual trapping, solubility trapping, and mineral trapping, and each of them plays a role at different times. Geomechanical issues directly determine the success or failure of GCS. The evolution of the in-situ stress and effective stress can be calculated from the pore pressure variation and the stress path coefficient. Further, the integrity of the caprock under different regime conditions can be evaluated based on the variation of in-situ stress. Generally speaking, it is most stable in the normal fault regime, followed by the strike-slip regime, and the most unstable is the thrust fault regime. In addition, acidic fluids or gases can be formed after CO_2 injection, corroding the wellbore and cement sheath, creating multiple leak paths, and leading to failure of wellbore integrity. For the general public, the top concern is whether the GCS project will cause earthquakes. Fortunately, it has been identified based on the current studies that the injection of CO_2 does trigger microseismic events that can be perceived by humans, but the magnitude of the earthquakes and the energy released would not bring damage to buildings or organisms on the ground.

DOI: http://dx.doi.org/10.5772/intechopen.105412

Author details

Yongcun Feng* and Shui Zhang
China University of Petroleum at Beijing, College of Petroleum Engineering, China

*Address all correspondence to: yongcun.f@gmail.com

References

[1] Adopted I. Climate Change 2014 Synthesis Report. IPCC: Geneva, Szwitzerland; 2014

[2] Wallquist L, Visschers VH, Siegrist M. Impact of knowledge and misconceptions on benefit and risk perception of CCS. Environmental Science & Technology. 2010;**44**:6557–6562

[3] Selma L, Seigo O, Dohle S, et al. Public perception of carbon capture and storage (CCS): A review. Renewable and Sustainable Energy Reviews. 2014;**38**: 848-863

[4] Gough C, Cunningham R, Mander S. Understanding key elements in establishing a social license for CCS: An empirical approach. International Journal of Greenhouse Gas Control. 2018;**68**:16-25

[5] Bai M. Risk assessment for CO_2 leakage along abandoned wells using a Monte Carlo simulation in a CO_2 sequestration site. Petroleum Science and Technology. 2014;**32**(10):1191-1200

[6] Zoback MD, Gorelick SM. Earthquake triggering and large-scale geologic storage of carbon dioxide. Proceedings of the National Academy of Sciences. 2012; **109**(26):10164-10168

[7] Metz B, Davidson O, De Coninck H, et al. IPCC Special Report on Carbon Dioxide Capture and Storage. Cambridge: Cambridge University Press; 2005

[8] Teng F, Tondeur D. Efficiency of carbon storage with leakage: Physical and economical approaches. Energy. 2007;**32**(4):540-548

[9] Klusman RW. Evaluation of leakage potential from a carbon dioxide EOR/ sequestration project. Energy Conversion and Management. 2003; **44**(12):1921-1940

[10] Saripalli KP, Mcgrail BP, White MD, et al. Modeling the sequestration of CO2 in deep geological formations. Laboratory directed research and development annual report-fiscal year. 2000;233-234

[11] Liteanu E, Spiers CJ, Peach CJ. Failure behaviour wellbore cement in the presence of water and supercritical CO_2. Energy Procedia. 2009;**1**(1):3553-3560

[12] Orlic B. Some geomechanical aspects of geological CO 2 sequestration. KSCE Journal of Civil Engineering. 2009;**13**(4): 225-232

[13] Gibbins J, Chalmers H. Carbon capture and storage. Energy Policy. 2008;**36**(12):4317-4322

[14] Shukla R, Ranjith P, Haque A, et al. A review of studies on CO_2 sequestration and caprock integrity. Fuel. 2010; **89**(10):2651-2664

[15] De Silva PNK, Ranjith P. A study of methodologies for CO_2 storage capacity estimation of saline aquifers. Fuel. 2012; **93**:13-27

[16] Iglauer S. Dissolution trapping of carbon dioxide in reservoir formation brine-a carbon storage mechanism. INTECH Open Access Publisher. 2011; 233-262

[17] Spiteri E, Juanes R, Blunt MJ, et al. Relative-permeability hysteresis: Trapping models and application to geological CO2 sequestration. In: Proceedings of the SPE Annual Technical Conference and Exhibition, Texas, USA: Society of Petroleum Engineers; 2005

[18] Gunter WD, Bachu S, Benson S. The role of hydrogeological and geochemical

trapping in sedimentary basins for secure geological storage of carbon dioxide. Geological Society, London, Special Publications. 2004;**233**(1): 129-145

[19] Kim Y, Jang H, Kim J, et al. Prediction of storage efficiency on CO_2 sequestration in deep saline aquifers using artificial neural network. Applied Energy. 2017;**185**:916-928

[20] Netl. CARBON STORAGE FAQS. U. S; National Energy Technology Laboratory. 2022. Available from: https://netl.doe.gov/coal/carbon-storage/faqs/carbon-storage-faqs

[21] Pentland C, El-Maghraby R, Georgiadis A, et al. Immiscible displacements and capillary trapping in CO_2 storage. Energy Procedia. 2011;**4**: 4969-4976

[22] Kneafsey TJ, Pruess K. Laboratory flow experiments for visualizing carbon dioxide-induced, density-driven brine convection. Transport in Porous Media. 2010;**82**(1):123-139

[23] Bachu S, Gunter W, Perkins E. Aquifer disposal of CO_2: Hydrodynamic and mineral trapping. Energy Conversion and Management. 1994; **35**(4):269-279

[24] Terzaghi K, Peck RB, Mesri G. Soil Mechanics in Engineering Practice. Canada: John Wiley & Sons; 1996;83-85

[25] Zhang JJ. Applied Petroleum Geomechanics. Texas, USA: Gulf Professional Publishing; 2019;5-8

[26] Das BM, Das B. Advanced Soil Mechanics. New York: Taylor & Francis. 2008;53-96

[27] Schutjens P, Hanssen T, Hettema M, et al. Compaction-induced porosity/ permeability reduction in sandstone reservoirs: Data and model for elasticity-dominated deformation. SPE Reservoir Evaluation & Engineering. 2004;**7**(03): 202-216

[28] Schutjens PM, Snippe JR, Mahani H, et al. Production-induced stress change in and above a reservoir pierced by two salt domes: A geomechanical model and its applications. SPE Journal. 2012; **17**(01):80-97

[29] Hettema M, Schutjens P, Verboom B, et al. Production-induced compaction of a sandstone reservoir: The strong influence of stress path. SPE Reservoir Evaluation & Engineering. 2000;**3**(04):342-347

[30] Gheibi S, Holt R, Vilarrasa V. Stress path evolution during fluid injection into geological formations. In: Proceedings of the 50th US Rock Mechanics/Geo-mechanics Symposium. Texas, USA; Society of Petroleum Engineers; 2016

[31] Lambe TW, Marr WA. Stress path method. Journal of the Geotechnical Engineering Division. 1979;**105**(6):727-738

[32] Parry RH. Mohr Circles, Stress Paths and Geotechnics. London: CRC Press; 2004;1-39

[33] Orlic B. Geomechanical effects of CO_2 storage in depleted gas reservoirs in the Netherlands: Inferences from feasibility studies and comparison with aquifer storage. Journal of Rock Mechanics and Geotechnical Engineering. 2016;**8**(6):846-859

[34] Gheibi S, Holt RM, Vilarrasa V. Effect of faults on stress path evolution during reservoir pressurization. International Journal of Greenhouse Gas Control. 2017;**63**:412-430

[35] Vilarrasa V, Makhnenko RY, Laloui L. Potential for fault reactivation

due to CO_2 injection in a semi-closed saline aquifer. Energy Procedia. 2017; **114**:3282-3290

[36] Cappa F, Rutqvist J. Impact of CO_2 geological sequestration on the nucleation of earthquakes. Geophysical Research Letters. 2011;**38**(17):1-6

[37] Rinaldi AP, Rutqvist J, Cappa F. Geomechanical effects on CO_2 leakage through fault zones during large-scale underground injection. International Journal of Greenhouse Gas Control. 2014;**20**:117-131

[38] Rutqvist J, Rinaldi AP, Cappa F, et al. Fault activation and induced seismicity in geological carbon storage–lessons learned from recent modeling studies. Journal of Rock Mechanics and Geotechnical Engineering. 2016;**8**(6): 789-804

[39] Paluszny A, Graham CC, Daniels KA, et al. Caprock integrity and public perception studies of carbon storage in depleted hydrocarbon reservoirs. International Journal of Greenhouse Gas Control. 2020;**98**: 103057

[40] Vilarrasa V. Impact of CO_2 injection through horizontal and vertical wells on the caprock mechanical stability. International Journal of Rock Mechanics and Mining Sciences. 2014;**66**:151-159

[41] De Simone S, Vilarrasa V, Carrera J, et al. Thermal coupling may control mechanical stability of geothermal reservoirs during cold water injection. Physics and Chemistry of the Earth, Parts A/B/C. 2013;**64**:117-126

[42] Vilarrasa V. The role of the stress regime on microseismicity induced by overpressure and cooling in geologic carbon storage. Geofluids. 2016;**16**(5): 941-953

[43] Vilarrasa V, Carrera J, Olivella S. Hydromechanical characterization of CO_2 injection sites. International Journal of Greenhouse Gas Control. 2013;**19**: 665-677

[44] Bachu S, Bennion DB. Experimental assessment of brine and/or CO_2 leakage through well cements at reservoir conditions. International Journal of Greenhouse Gas Control. 2009;**3**(4): 494-501

[45] Viswanathan HS, Pawar RJ, Stauffer PH, et al. Development of a hybrid process and system model for the assessment of wellbore leakage at a geologic CO_2 sequestration site. Environmental Science & Technology. 2008;**42**(19):7280-7286

[46] Carroll S, Carey JW, Dzombak D, et al. Role of chemistry, mechanics, and transport on well integrity in CO_2 storage environments. International Journal of Greenhouse Gas Control. 2016;**49**:149-160

[47] Reinicke K, Fichter C. Measurement strategies to evaluate the integrity of deep wells for CO_2 applications. Underground Storage of CO_2 and Energy. 2010:67-74

[48] Zhang Y, Gao K, Schmitt G. Inhibition of steel corrosion under aqueous supercritical CO_2 conditions. In: Proceedings of the CORROSION 2011, Texas, USA: Society of Petroleum Engineers; 2011

[49] Kermani M, Morshed A. Carbon dioxide corrosion in oil and gas productiona compendium. Corrosion. 2003;**59**(08)

[50] Carey JW, Wigand M, Chipera SJ, et al. Analysis and performance of oil well cement with 30 years of CO_2 exposure from the SACROC unit, West Texas, USA. International Journal of

Greenhouse Gas Control. 2007;**1**(1): 75-85

[51] Kutchko BG, Strazisar BR, Dzombak DA, et al. Degradation of well cement by CO_2 under geologic sequestration conditions. Environmental Science & Technology. 2007;**41**(13): 4787-4792

[52] Gray KE, Podnos E, Becker E. Finite-element studies of near-wellbore region during cementing operations: Part I. SPE Drilling & Completion. 2009;**24**(01): 127-136

[53] Zhang M, Bachu S. Review of integrity of existing wells in relation to CO_2 geological storage: What do we know? International Journal of Greenhouse Gas Control. 2011;**5**(4): 826-840

[54] Ellsworth WL. Injection-induced earthquakes. Science. 2013;**341**:6142

[55] Grigoli F, Cesca S, Rinaldi AP, et al. The November 2017 mw 5.5 Pohang earthquake: A possible case of induced seismicity in South Korea. Science. 2018; **360**(6392):1003-1006

[56] Council NR. Induced Seismicity Potential in Energy Technologies. Washington, USA: National Academies Press; 2013

[57] Verdon J, Kendall J-M, White D, et al. Linking microseismic event observations with geomechanical models to minimise the risks of storing CO_2 in geological formations. Earth and Planetary Science Letters. 2011;**305**(1–2): 143-152

[58] Paterson L, Lu M, Connell L, et al. Numerical modeling of pressure and temperature profiles including phase transitions in carbon dioxide wells. In: Proceedings of the SPE Annual Technical Conference and Exhibition, Denver, USA; Society of Petroleum Engineers; 2008

[59] Vilarrasa V, Rutqvist J. Thermal effects on geologic carbon storage. Earth-Science Reviews. 2017;**165**:245-256

[60] Mcgarr A. Maximum magnitude earthquakes induced by fluid injection. Journal of Geophysical Research Solid Earth. 2014:**119**(2):1008-1019